Tractatus Logico-Philosophicus

Start Publishing PD LLC
Copyright © 2024 by Start Publishing PD LLC

All rights reserved, including the right to reproduce this book or portions thereof in any form whatsoever.

Start Publishing PD is a registered trademark of Start Publishing PD LLC
Manufactured in the United States of America

Cover art: Shutterstock/Taisiya Kozorez

Cover design: Jennifer Do

10 9 8 7 6 5 4 3 2 1

ISBN 979-8-8809-2399-1

Tractatus Logico-Philosophicus
Complete and Unabridged
by Ludwig Wittgenstein

Introduction
By Bertrand Russell

Mr Wittgenstein's Tractatus Logico-Philosophicus, whether or not it prove to give the ultimate truth on the matters with which it deals, certainly deserves, by its breadth and scope and profundity, to be considered an important event in the philosophical world. Starting from the principles of Symbolism and the relations which are necessary between words and things in any language, it applies the result of this inquiry to various departments of traditional philosophy, showing in each case how traditional philosophy and traditional solutions arise out of ignorance of the principles of Symbolism and out of misuse of language.

The logical structure of propositions and the nature of logical inference are first dealt with. Thence we pass successively to Theory of Knowledge, Principles of Physics, Ethics, and finally to the Mystical (das Mystische).

In order to understand Mr Wittgenstein's book, it is necessary to realize what is the problem with which he is concerned. In the part of his theory which deals with Symbolism he is concerned with the conditions which would have to be fulfilled by a logically perfect language. There are various problems as regards language. First, there is the problem what actually occurs in our minds when we use language with the intention of meaning something by it; this problem belongs to psychology. Secondly, there is the problem as to what is the relation subsisting between thoughts, words, or sentences, and that which they refer to or mean; this problem belongs to epistemology. Thirdly, there is the problem of using sentences so as to convey truth rather that falsehood; this belongs to the special sciences dealing with the subject-matter of the sentences in question. Fourthly, there is the question: what relation must one fact (such as a sentence) have to another in order to be capable of being a symbol for that other? This last is a logical question, and is the one with which Mr Wittgenstein is concerned. He is concerned with the conditions for accurate Symbolism, i.e. for Symbolism in which a sentence 'means' something quite definite. In practice, language is always more or less vague, so that what we assert is never quite precise. Thus, logic has two problems to deal with in regard to Symbolism: (1) the conditions for sense rather than nonsense in combinations of words; (2) the conditions for uniqueness of meaning or reference in symbols or combinations of symbols. A logically perfect language has rules of syntax which prevent nonsense, and has single symbols which always have a definite and unique meaning. Mr Wittgenstein is concerned with the conditions for a logically perfect language — not that any language is logically perfect, or that we believe ourselves capable, here and now, of constructing a logically perfect language, but that the whole function of language is to have meaning, and it only fulfills this function in proportion as it approaches to the ideal language which we postulate.

The essential business of language is to assert or deny facts. Given the syntax of language, the meaning of a sentence is determined as soon as the meaning of the component words is known. In order that a certain sentence should assert a certain fact there must, however the language may be constructed, be something in common between the structure of the sentence and the structure of the fact. This is perhaps the most fundamental thesis of Mr Wittgenstein's theory. That which has to be in common between the sentence and the fact cannot, he contends, be itself in turn said in language. It can, in his phraseology, only be shown, not said, for whatever we may say will still need to have the same structure.

The first requisite of an ideal language would be that there should be one name for every simple, and never the same name for two different simples. A name is a simple symbol in the sense that it has no parts which are themselves symbols. In a logically perfect language nothing that is not simple will have a simple symbol. The symbol for the whole will be a "complex", containing the symbols for the parts. In speaking of a "complex" we are, as will appear later, sinning against the rules of philosophical grammar, but this is unavoidable at the outset. "Most propositions and questions that have been written about philosophical matters are not false but senseless. We cannot, therefore, answer questions of this kind at all, but only state their senselessness. Most questions and propositions of the philosopohers result from the fact that we do not understand the logic of our language. They are of the same kind as the question whether the Good is more or less identical than the Beautiful" (4.003). What is complex in the world is a fact. Facts which are not compounded of other facts are what Mr Wittgenstein calls Sachverhalte, whereas a fact which may consist of two or more facts is a Tatsache: thus, for example "Socrates is wise" is a Sachverhalt, as well as a Tatsache, whereas "Socrates is wise and Plato is his pupil" is a Tatsache but not a Sachverhalt.

He compares linguistic expression to projection in geometry. A geometrical figure may be projected in many ways: each of these ways corresponds to a different language, but the projective properties of the original figure remain unchanged whichever of these ways may be adopted. These projective properties correspond to that which in his theory the proposition and the fact must have in common, if the proposition is to assert the fact.

In certain elementary ways this is, of course, obvious. It is impossible, for example, to make a statement about two men (assuming for the moment that the men may be treated as simples), without employing two names, and if you are going to assert a relation between the two men it will be necessary that the sentence in which you make the assertion shall establish a relation between the two names. If we say "Plato loves Socrates", the word "loves" which occurs between the word "Plato" and the word "Socrates" establishes a certain relation between these two words, and it is owing to this fact that our sentence is able to assert a relation between the persons named by the words "Plato" and "Socrates". "We must not say, the complex sign 'aRb' says that 'a stands in a certain relation R to b'; but we must say, that 'a' stands in a certain relation to 'b' says that aRb' (3.1432).

Mr Wittgenstein begins his theory of Symbolism with the statement (2.1): "We make to ourselves pictures of facts." A picture, he says, is a model of the reality, and to the objects in the reality correspond the elements of the picture: the picture itself is a fact. The fact that things have a certain relation to each other is represented by the fact that in the picture its elements have a certain relation to one another. "In the picture and the pictured there must be something identical in order that the one can be a picture of the other at all. What the picture must have in common with reality in order to be able to represent it after its manner — rightly or falsely — is its form of representation" (2.161, 2.17).

We speak of a logical picture of a reality when we wish to imply only so much resemblance as is essential to its being a picture in any sense, that is to say, when we wish to imply no more than identity of logical form. The logical picture of a fact, he says, is a Gedanke. A picture can correspond or not correspond with the fact and be accordingly true or false, but in both cases it shares the logical form with the fact. The sense in which he speaks of pictures is illustrated by his statement: "The gramophone record, the musical thought, the score, the waves of sound, all stand to one another in that pictorial internal relation which holds between language and the world. To all of them the logical structure is common. (Like the two youths, their two horses and their lilies in the story. They are all in a certain sense one)" (4.014). The possibility of a proposition representing a fact rests upon the fact that in it objects are represented by signs, but are themselves present in the proposition as in the fact. The proposition and the fact must exhibit the same logical "manifold," and this cannot be itself represented since it has to be in common between the fact and the picture. Mr Wittgenstein maintains that everything properly philosophical belongs to what can only be shown, or to what is in common between a fact and its logical picture. It results from this view that nothing correct can be said in philosophy. Every philosophical proposition is bad grammar, and the best that we can hope to achieve by philosophical discussion is to lead people to see that philosophical discussion is a mistake. "Philosophy is not one of the natural sciences. (The word 'philosophy' must mean something which stands above or below, but not beside the natural sciences.) The object of philosophy is the logical clarification of thoughts. Philosophy is not a theory but an activity. A philosophical work consists essentially of elucidations. The result of philosophy is not a number of 'philosophical propositions,' but to make propositions clear. Philosophy should make clear and delimit sharply the thoughts which otherwise are, as it were, opaque and blurred" (4.111 and 4.112). In accordance with this principle the things that have to be said in leading the reader to understand Mr Wittgenstein's theory are all of them things which that theory itself condemns as meaningless. With this proviso we will endeavour to convey the picture of the world which seems to underlie his system.

The world consists of facts: facts cannot strictly speaking be defined, but we can explain what we mean by saying that facts are what makes propositions true, or false. Facts may contain parts which are facts or may contain no such parts; for example: "Socrates was a wise Athenian," consists of the two facts, "Socrates was wise," and "Socrates was an Athenian." A fact which has no parts that are facts is called by Mr Wittgenstein a Sachverhalt. This is the same thing that he calls an atomic fact. An atomic fact, although it contains no parts that are facts, nevertheless does contain parts. If we may regard "Socrates is wise" as an atomic fact we perceive that it contains the constituents "Socrates" and "wise." If an atomic fact is analyzed as fully as possibly (theoretical, not practical possibility is meant) the constituents finally reached may be called "simples" or "objects." It is a logical necessity demanded by theory, like an electron. His ground for maintaining that there must be simples is that every complex presupposes a fact. It is not necessarily assumed that the complexity of facts is finite; even if every fact consisted of an infinite number of atomic facts and if every atomic fact consisted of an infinite number of objects there would still be objects and atomic facts (4.2211). The assertion that there is a certain complex reduces to the assertion that its constituents are related in a certain way, which is the assertion of a fact: thus if we give a name to the complex the name only has meaning in virtue of the truth of a certain proposition, namely the proposition asserting the relatedness of the constituents of the complex. Thus the naming of complexes presupposes propositions, while propositions presupposes the naming of simples. In this way the naming of simples is shown to be what is logically first in logic.

The world is fully described if all atomic facts are known, together with the fact that these are all of them. The world is not described by mearly naming all the objects in it; it is necessary also to know the atomic facts of which these objects are constituents. Given this total of atomic facts, every true proposition, however complex, can theoretically be inferred. A proposition (true or false) asserting an atomic fact is called an atomic proposition. All atomic propositions are logically independent of each other. No atomic proposition implies any other or is inconsistent with any other. Thus the whole business of logical inference is concerned with proposition which are not atomic. Such propositions may be called molecular.

Wittgenstein's theory of molecular propositions turns upon his theory of the construction of truth-functions.

A truth-function of a proposition p is a proposition containing p and such that its truth or falsehood depends only upon the truth or falsehood of p, and similarly a truth-function of several propositions p, q, r, . . . is one containing p, q, r . . . and such that its truth or falsehood depends only upon the truth or falsehood of p, q, r . . . . It might seem at first sight as though there were other functions of propositions besides truth-functions; such, for example, would be "A believes p," for in general A will believe some true propositions and some false ones: unless he is an exceptionally gifted individual, we cannot infer that p is true from the fact that he believes it or that p is false from the fact that he does not believe it. Other apparent exceptions would be such as "p is a very complex proposition" or "p is a proposition about Socrates." Mr Wittgenstein maintains, however, for reasons which will appear presently, that such exceptions are only apparent, and that every function of a proposition is really a truth-function. It follows that if we can define truth-functions generally, we can obtain a general definition of all propositions in terms of the original set of atomic propositions. This Wittgenstein proceeds to do.

It has been shown by Dr. Sheffer (Trans. Am. Math. Soc., Vol. XIV. pp. 481-488) that all truth-functions of a given set of propositions can be constructed out of either of the two functions "not-p or not-q" or "not-p and not-q". Wittgenstein makes use of the latter, assuming a knowledge of Dr. Sheffer's work. The manner in which other truth-functions are constructed out of "not-p and not-q" is easy to see. "Not-p and not-p" is equivalent to "not-p," hence we obtain a definition of negation in terms of our primitive function: hence we can define "p or q," since this is the negation of "not-p and not-q," i.e. of our primitive function. The development of other truth-functions out of "not-p" and "p or q" is given in detail at the beginning of Principia Mathematica. This gives all that is wanted when the propositions which are arguments to our truth-function are given by enumeration. Wittgenstein, however, by a very interesting analysis succeeds in extending the process to general propositions, i.e. to cases where the propositions which are arguments to our truth-function are not given by enumeration but are given as all those satisfying some condition. For example, let fx be a propositional function (i.e. a function whose values are propositions), such as "x is human" — then the various values of fx form a set of propositions. We may extend the idea "not-p and not-q" so as to apply to simultaneous denial of all the propositions which are values of fx. In this way we arrive at the proposition which is ordinarily represented in mathematical logic by the words "fx is false for all values of x." The negation of this would be the proposition "There is at least one x for which fx is true" which is represented by "(EXISTS x) . fx." If we had started with not-fx instead of fx we should have arrived at the proposition "fx is true for all values of x" which is represented by "(x) .fx. " Wittgenstein's method of dealing with general propositions [i.e. "(x) . fx" and "(EXISTS x) .fx"] differs from previous methods by the fact that the generality comes only in specifying the set of propositions concerned, and when this has been done the building up of truth-functions proceeds exactly as it would in the case of a finite number of enumerated arguments p, q, r . . . .

 Mr Wittgenstein's explanation of his symbolism at this point is not quite fully given in the text. The symbol he uses is [p-bar , xi-bar , N(xi-bar)]. The following is the explanation of this symbol:

p-bar stands for all atomic propositions.

xi-bar stands for any set of propositions.

N(xi-bar) stands for the negation of all the propositions making up xi .

The whole symbol [p-bar , xi-bar , N(xi-bar)] means whatever can be obtained by taking any selection of atomic propositions, negating them all, then taking any selection of the set of propositions now obtained, together with any of the originals — and so on indefinitely. This is, he says, the general truth-function and also the general form of proposition. What is meant is somewhat less complicated than it sounds. The symbol is intended to describe a process by the help of which, given the atomic propositions, all others can be manufactured. The process depends upon:

(a) Sheffer's proof that all truth-functions can be obtained out of simultaneous negation, i.e. out of "not-p and not-q";

(b) Mr Wittgenstein's theory of the derivation of general propositions from conjunctions and disjunctions;

(c) The assertion that a proposition can only occur in another proposition as argument to a truth-function. Given these three foundations, it follows that all propositions which are not atomic can be derived from such as are, buy a uniform process, and it is this process which is indicated by Mr Wittgenstein's symbol.

From this uniform method of construction we arrive at an amazing simplification of the theory of inference, as well as a definition of the sort of propositions that belong to logic. The method of generation which has just been described, enables Wittgenstein to say that all propositions can be constructed in the above manner from atomic propositions, and in this way the totality of propositions is defined. (The apparent exceptions which we mentioned above are delt with in a manner which we shall consider later.) Wittgenstein is enabled to assert that propositions are all that follows from the totality of atomic propositions (together with the fact that it is the totality of them); that a proposition is always a truth-function of atomic propositions; and that if p follows from q the meaning of p is contained in the meaning of q, from which of course it results that nothing can be deduced from an atomic proposition. All the propositions of logic, he maintains, are tautologies, such, for example, as "p or not p."

The fact that nothing can be deduced from an atomic proposition has interesting applications, for example, to causality. There cannot, in Wittgenstein's logic, be any such thing as a causal nexus. "The events of the future," he says, "cannot be inferred from those of the present. Superstition is the belief in the causal nexus." That the sun will rise to-morrow is a hypothesis. We do not in fact know whether it will rise, since there is no compulsion according to which one thing must happen because another happens.

Let us now take up another subject — that of names. In Wittgenstein's theoretical logical language, names are only given to simples. We do not give two names to one thing, or one name to two things. There is no way whatever, according to him, by which we can describe the totality of things that can be names, in other words, the totality of what there is in the world. In order to be able to do this we should have to know of some property which must belong to every thing by a logical necessity. It has been sought to find such a property in self-identity, but the conception of identity is subjected by Wittgenstein to a destructive criticism from which there seems no escape. The definition of identity by means of the identity of indiscernibles is rejected, because the identity of indiscernibles appears to be not a logically necessary principle. According to this principle x is identical with y if every property of x is a property of y; but it would, after all be logically possible for two things to have exactly the same properties. If this does not in fact happen that is an accidental characteristic of the world, not a logically necessary characteristic, and accidental characteristics of the world must, of course, not be admitted into the structure of logic. Mr Wittgenstein accordingly banishes identity and adopts the convention that different letters are to mean different things. In practice, identity is needed as between a name and a description or between two descriptions. It is needed for such propositions as "Socrates is the philosopher who drank the hemlock," or "The even prime is the next number after 1." For such uses of identity it is easy to provide on Wittgenstein's system.

The rejection of identity removes one method of speaking of the totality of things, and it will be found that any other method that may be suggested is equally fallacious: so, at least, Wittgenstein contends and, I think, rightly. This amounts to saying that "object" is a pseudo-concept. To say "x is an object" is to say nothing. It follows from this that we cannot make such statements as "there are more than three objects in the world," or "there are an infinite number of objects in the world." Objects can only be mentioned in connexion with some definite property. We can say "there are more than three objects which are human," or "there are more than three objects which are red," for in these statements the word object can be replaced by a variable in the language of logic, the variable being one which satisfies in the first case the function "x is human"; in the second the function "x is red." But when we attempt to say "there are more than three objects," this substitution of the variable for the word "object" becomes impossible, and the proposition is therefore seen to be meaningless.

We here touch one instance of Wittgenstein's fundamental thesis, that it is impossible to say anything about the world as a whole, and that whatever can be said has to be about bounded portions of the world. This view may have been originally suggested by notation, and if so, that is much in its favor, for a good notation has a subtlety and suggestiveness which at times make it seem almost like a live teacher. Notational irregularities are often the first sign of philosophical errors, and a perfect notation would be a substitute for thought. But although notation may have first suggested to Mr Wittgenstein the limitation of logic to things within the world as opposed to the world as a whole, yet the view, once suggested, is seen to have much else to recommend it. Whether it is ultimately true I do not, for my part, profess to know. In this Introduction I am concerned to expound it, not to pronounce upon it. According to this view we could only say things about the world as a whole if we could get outside the world, if, that is to say, it ceased to be for us the whole world. Our world may be bounded for some superior being who can survey it from above, but for us, however finite it may be, it cannot have a boundary, since it has nothing outside it. Wittgenstein uses, as an analogy, the field of vision. Our field of vision does not, for us, have a visual boundary, just because there is nothing outside it, and in like manner our logical world has no logical boundary because our logic knows of nothing outside it. These considerations lead him to a somewhat curious discussion of Solipsism. Logic, he says, fills the world. The boundaries of the world are also its boundaries. In logic, therefore, we cannot say, there is this and this in the world, but not that, for to say so would apparently presuppose that we exclude certain possibilities, and this cannot be the case, since it would require that logic should go beyond the boundaries of the world as if it could contemplate these boundaries from the other side also. What we cannot think we cannot think, therefore we also cannot say what we cannot think.

This, he says, gives the key to solipsism. What Solipsism intends is quite correct, but this cannot be said, it can only be shown. That the world is my world appears in the fact that the boundaries of language (the only language I understand) indicate the boundaries of my world. The metaphysical subject does not belong to the world but is a boundary of the world.

We must take up next the question of molecular propositions which are at first sight not truth-functions, of the propositions that they contain, such, for example, as "A believe p."

Wittgenstein introduces this subject in the statement of his position, namely, that all molecular functions are truth-functions. He says (5.54): "In the general propositional form, propositions occur in a proposition only as bases of truth-operations." At first sight, he goes on to explain, it seems as if a propositions could also occur in other ways, e.g. "A believes p." Here it seems superficially as if the proposition p stood in a sort of relation to the object A. "But it is clear that 'A believe that p,' 'A thinks p,' 'A says p' are of the form enumerated arguments p, q, r . . .'p says p'; and here we have no co-ordination of a fact and an object, but a co-ordination of facts by means of a co-ordination of their objects" (5.542).

What Mr Wittgenstein says here is said so shortly that its point is not likely to be clear to those who have not in mind the controversies with which he is concerned. The theory which which he is disagreeing will be found in my articles on the nature of truth and falsehood in Philosophical Essays and Proceedings of the Aristotelian Society, 1906-7. The problem at issue is the problem of the logical form of belief, i.e. what is the schema representing what occurs when a man believes. Of course, the problem applies not only to belief, but also to a host of other mental phenomena which may be called propositional attitudes: doubting, considering, desiring, etc. In all these cases it seems natural to express the phenomenon in the form "A doubts p," "A desires p," etc., which makes it appear as though we were dealing with a relation between a person and a proposition. This cannot, of course, be the ultimate analysis, since persons are fictions and so are propositions, except in the sense in which they are facts on their own account. A proposition, considered as a fact on its own account, may be a set of words which a man says over to himself, or a complex image, or train of images passing through his mind, or a set of incipient bodily movements. It may be any one of innumerable different things. The proposition as a fact on its own account, for example the actual set of words the man pronounces to himself, is not relevant to logic. What is relevant to logic is that common element among all these facts, which enables him, as we say, to mean the fact which the proposition asserts. To psychology, of course, more is relevant; for a symbol does not mean what it symbolizes in virtue of a logical relation alone, but in virtue also of a psychological relation of intention, or association or what-not. The psychological part of meaning, however, does not concern the logician. What does concern him in this problem of belief is the logical schema. It is clear that, when a person believes a proposition, the person, considered as a metaphysical subject, does not have to be assumed in order to explain what is happening. What has to be explained is the relation between the set of words which is the proposition considered as a fact on its own account, and the "objective" fact which makes the proposition true or false. This reduces ultimately to the question of the meaning of propositions, that is to say, the meaning of propositions is the only non-psychological portion of the problem involved in the analysis of belief. This problem is simply one of a relation of two facts, namely, the relation between the series of words used by the believer and the fact which makes these words true or false. The serious of words is a fact just as much as what makes it true or false is a fact. The relation between these two facts is not unanalyzable, since the meaning of a proposition results from the meaning of its constituent words. The meaning of the series of words which is a proposition is a function of the meaning of the separate words. Accordingly, the proposition as a whole does not really enter into what has to be explained in explaining the meaning of a propositions. It would perhaps help to suggest the point of view which I am trying to indicate, to say that in the cases which have been considering the proposition occurs as a fact, not as a proposition. Such a statement, however, must not be taken too literally. The real point is that in believing, desiring, etc., what is logically fundamental is the relation of a proposition considered as a fact to the fact which makes it true or false, and that this relation of two facts is reducible to a relation of their constituents. Thus the proposition does not occur at all in the same sense in which it occurs in a truth-function.

There are some respects, in which, as it seems to me, Mr Wittgenstein's theory stands in need of greater technical development. This applies in particular to his theory of number (6.02 ff.) which, as it stands, is only capable of dealing with finite numbers. No logic can be considered adequate until it has been shown to be capable of dealing with transfinite numbers. I do not think there is anything in Mr Wittgenstein's system to make it impossible for him to fill this lacuna.

More interesting than such questions of comparative detail is Mr Wittgenstein's attitude towards the mystical. His attitude upon this grows naturally out of his doctrine in pure logic, according to which the logical proposition is a picture (true or false) of the fact, and has in common with the fact a certain structure. It is this common structure which makes it capable of being a picture of the fact, but the structure cannot itself be put into words, since it is a structure of words, as well as of the fact to which they refer. Everything, therefore, which is involved in the very idea of the expressiveness of language must remain incapable of being expressed in language, and is, therefore, inexpressible in a perfectly precise sense. This inexpressible contains, according to Mr Wittgenstein, the whole of logic and philosophy, he says, would be to confine oneself to propositions of the sciences, stated with all possible clearness and exactness, leaving philosophical assertions to the learner, and proving to him, whenever he made them, that they are meaningless. It is true that the fate of Socrates might befall a man who attempted this method of teaching, but we are not to be deterred by that fear, if it is the only right method. It is not this that causes some hesitation in accepting Mr Wittgenstein's position, in spite of the very powerful arguments which he brings to its support. What causes hesitation is the fact that, after all, Mr Wittgenstein manages to say a good deal about what cannot be said, thus suggesting to the sceptical reader that possibly there may be some loophole through a hierarchy of languages, or by some other exit. The whole subject of ethics, for example, is placed by Mr Wittgenstein in the mystical, inexpressible region. Nevertheless he is capable of conveying his ethical opinions. His defence would be that what he calls the mystical can be shown, although it cannot be said. It may be that this defence is adequate, but, for my part, I confess that it leaves me with acertain sense of intellectual discomfort.

There is one purely logical problem in regard to which these difficulties are peculiarly acute. I mean the problem of generality. In the theory of generality it is necessary to consider all propositions of the form fx where fx is a given propositional functions. This belongs to the part of logic which can be expressed, according to Mr Wittgenstein's system. But the totality of possible values of x which might seem to be involved in the totality of propositions of the form fx is not admitted by Mr Wittgenstein among the things that can be spoken of, for this is no other than the totality of things in the world, and thus involves the attempt to conceive the world as a whole; "the feeling of the world as a bounded whole is the mystical"; hence the totality of the values of x is mystical (6.45). This is expressly argued when Mr Wittgenstein denies that we can make propositions as to how may things there are in the world, as for example, that there are more than three.

These difficulties suggest to my mind some such possibility as this: that every language has, as Mr Wittgenstein says, a structure concerning which in the language, nothing can be said, but that there may be another language dealing with the structure of the first language, and having itself a new structure, and that to this hierarchy of languages there may be no limit. Mr Wittgenstein would of course reply that his whole theory is applicable unchanged to the totality of such languages. The only retort would be to deny that there is any such totality. The totalities concerning which Mr Wittgenstein holds that it is impossible to speak logically are nevertheless thought by him to exist, and are the subject-matter of his mysticism. The totality resulting from our hierarchy would be not merely logically inexpressible, but a fiction, a mere delusion, and in this way the supposed sphere of the mystical would be abolished. Such a hypothesis is very difficult, and I can see objections to it which at the moment I do not know how to answer. Yet I do not see how any easier hypothesis can escape from Mr Wittgenstein's conclusions. Even if this very difficult hypothesis should prove tenable, it would leave untouched a very large part of Mr Wittgenstein's theory, though possibly not the part upon which he himself would wish to lay most stress. As one with a long experience of the difficulties of logic and of the deceptiveness of theories which seem irrefutable, I find myself unable to be sure of the rightness of a theory, mearly on the ground that I cannot see any point on which it is wrong. But to have constructed a theory of logic which is not at any point obviously wrong is to have achieved a work of extraordinary difficulty and importance. This merit, in my opinion, belongs to Mr Wittgenstein's book, and makes it one which no serious philosopher can afford to neglect.

Bertrand Russell

May 1922

Preface

This book will perhaps only be understood by those who have themselves already thought the thoughts which are expressed in it — or similar thoughts. it is therefore not a text-book. Its object would be attained if it afforded pleasure to one who read it with understanding.

The book deals with the problems of philosophy and shows, as I believe, that the method of formulating these problems rests on the misunderstanding of the logic of our language. Its whole meaning could be summed up somewhat as follows: What can be said at all can be said clearly; and whereof one cannot speak thereof one must be silent.

The book will, therefore, draw a limit to thinking, or rather — not to thinking, but to the expression of thoughts; for, in order to draw a limit to thinking we should have to be able to thnk both sides of this limit (we should therefore have to be able to think what cannot be thought).

The limit can, therefore, only be drawn in language and what lies on the other side of the limit will be simply nonsense.

How far my efforts agree with those of other philosophers I will not decide. Indeed what I have here written makes no claim to novelty in points of detail; and therefore I give no sources, because it is indifferent to me whether what I have thought has already been thought before my by another.

I will only mention that to the great works of Frege and the writings of my friend Bertrand Russell I owe in large measure the stimulation of my thoughts.

If this work has a value it consists in two things. First that in it thoughts are expressed, and this value will be the greater the better the thoughts are expressed. the more the nail has been hit on the heard. — Here I am conscious that I have fallen far short of the possible. Simply because my powers are insufficient to cope with the task. — May others come and do it better.

On the other hand the truth of the thoughts communicated here seems to me unassailable and definitive. I am, therefore, of the opinion that the problems have in essentials been finally solved. And if I am not mistaken in this, then the value of this work secondly consists in the fact that it shows how little has been done when these problems have been solved.

L. W.

Vienna, 1918

Tractatus Logico-Philosophicus

1 The world is all that is the case.

1.1 The world is the totality of facts, not of things.

1.11 The world is determined by the facts, and by their being all the facts.

1.12 For the totality of facts determines what is the case, and also whatever is not the case.

1.13 The facts in logical space are the world.

1.2 The world divides into facts.

1.21 Each item can be the case or not the case while everything else remains the same.

2 What is the case—a fact—is the existence of states of affairs.

2.01 A state of affairs (a state of things) is a combination of objects (things).

2.011 It is essential to things that they should be possible constituents of states of affairs.

2.012 In logic nothing is accidental: if a thing can occur in a state of affairs, the possibility of the state of affairs must be written into the thing itself.

2.0121 It would seem to be a sort of accident, if it turned out that a situation would fit a thing that could already exist entirely on its own. If things can occur in states of affairs, this possibility must be in them from the beginning. (Nothing in the province of logic can be merely possible. Logic deals with every possibility and all possibilities are its facts.) Just as we are quite unable to imagine spatial objects outside space or temporal objects outside time, so too there is no object that we can imagine excluded from the possibility of combining with others. If I can imagine objects combined in states of affairs, I cannot imagine them excluded from the possibility of such combinations.

2.0122 Things are independent in so far as they can occur in all possible situations, but this form of independence is a form of connexion with states of affairs, a form of dependence. (It is impossible for words to appear in two different roles: by themselves, and in propositions.)

2.0123 If I know an object I also know all its possible occurrences in states of affairs. (Every one of these possibilities must be part of the nature of the object.) A new possibility cannot be discovered later.

2.01231 If I am to know an object, thought I need not know its external properties, I must know all its internal properties.

2.0124 If all objects are given, then at the same time all possible states of affairs are also given.

2.013 Each thing is, as it were, in a space of possible states of affairs. This space I can imagine empty, but I cannot imagine the thing without the space.

2.0131 A spatial object must be situated in infinite space. (A spatial point is an argument-place.) A speck in the visual field, thought it need not be red, must have some colour: it is, so to speak, surrounded by . Notes must have some pitch, objects of the sense of touch some degree of hardness, and so on.

2.014 Objects contain the possibility of all situations.

2.0141 The possibility of its occurring in states of affairs is the form of an object.

2.02 Objects are simple.

2.0201 Every statement about complexes can be resolved into a statement about their constituents and into the propositions that describe the complexes completely.

2.021 Objects make up the substance of the world. That is why they cannot be composite.

2.0211 If they world had no substance, then whether a proposition had sense would depend on whether another proposition was true.

2.0212 In that case we could not sketch any picture of the world (true or false).

2.022 It is obvious that an imagined world, however difference it may be from the real one, must have something—a form—in common with it.

2.023 Objects are just what constitute this unalterable form.

2.0231 The substance of the world can only determine a form, and not any material properties. For it is only by means of propositions that material properties are represented—only by the configuration of objects that they are produced.

2.0232 In a manner of speaking, objects are colourless.

2.0233 If two objects have the same logical form, the only distinction between them, apart from their external properties, is that they are different.

2.02331 Either a thing has properties that nothing else has, in which case we can immediately use a description to distinguish it from the others and refer to it; or, on the other hand, there are several things that have the whole set of their properties in common, in which case it is quite impossible to indicate one of them. For it there is nothing to distinguish a thing, I cannot distinguish it, since otherwise it would be distinguished after all.

2.024 The substance is what subsists independently of what is the case.

2.025 It is form and content.

2.0251 Space, time, colour (being coloured) are forms of objects.

2.026 There must be objects, if the world is to have unalterable form.

2.027 Objects, the unalterable, and the subsistent are one and the same.

2.0271 Objects are what is unalterable and subsistent; their configuration is what is changing and unstable.

2.0272 The configuration of objects produces states of affairs.

2.03 In a state of affairs objects fit into one another like the links of a chain.

2.031 In a state of affairs objects stand in a determinate relation to one another.

2.032 The determinate way in which objects are connected in a state of affairs is the structure of the state of affairs.

2.033 Form is the possibility of structure.

2.034 The structure of a fact consists of the structures of states of affairs.

2.04 The totality of existing states of affairs is the world.

2.05 The totality of existing states of affairs also determines which states of affairs do not exist.

2.06 The existence and non-existence of states of affairs is reality. (We call the existence of states of affairs a positive fact, and their a negative fact.)

2.061 States of affairs are independent of one another.

2.062 From the existence or non-existence of one state of affairs it is impossible to infer the existence or non-existence of another.

2.063 The sum-total of reality is the world.

2.1 We picture facts to ourselves.

2.11 A picture presents a situation in logical space, the existence and of states of affairs.

2.12 A picture is a model of reality.

2.13 In a picture objects have the elements of the picture corresponding to them.

2.131 In a picture the elements of the picture are the representatives of objects.

2.14 What constitutes a picture is that its elements are related to one another in a determinate way.

2.141 A picture is a fact.

2.15 The fact that the elements of a picture are related to one another in a determinate way represents that things are related to one another in the same way. Let us call this connexion of its elements the structure of the picture, and let us call the possibility of this structure the pictorial form of the picture.

2.151 Pictorial form is the possibility that things are related to one another in the same way as the elements of the picture.

2.1511 That is how a picture is attached to reality; it reaches right out to it.

2.1512 It is laid against reality like a measure.

2.15121 Only the end-points of the graduating lines actually touch the object that is to be measured.

2.1514 So a picture, conceived in this way, also includes the pictorial relationship, which makes it into a picture.

2.1515 These correlations are, as it were, the feelers of the picture's elements, with which the picture touches reality.

2.16 If a fact is to be a picture, it must have something in common with what it depicts.

2.161 There must be something identical in a picture and what it depicts, to enable the one to be a picture of the other at all.

2.17 What a picture must have in common with reality, in order to be able to depict it—correctly or incorrectly—in the way that it does, is its pictorial form.

2.171 A picture can depict any reality whose form it has. A spatial picture can depict anything spatial, a coloured one anything coloured, etc.

2.172 A picture cannot, however, depict its pictorial form: it displays it.

2.173 A picture represents its subject from a position outside it. (Its standpoint is its representational form.) That is why a picture represents its subject correctly or incorrectly.

2.174 A picture cannot, however, place itself outside its representational form.

2.18 What any picture, of whatever form, must have in common with reality, in order to be able to depict it—correctly or incorrectly—in any way at all, is logical form, i.e. the form of reality.

2.181 A picture whose pictorial form is logical form is called a logical picture.

2.182 Every picture is at the same time a logical one. (On the other hand, not every picture is, for example, a spatial one.)

2.19 Logical pictures can depict the world.

2.2 A picture has logico-pictorial form in common with what it depicts.

2.201 A picture depicts reality by representing a possibility of existence and non-existence of states of affairs.

2.202 A picture contains the possibility of the situation that it represents.

2.203 A picture agrees with reality or fails to agree; it is correct or incorrect, true or false.

2.22 What a picture represents it represents independently of its truth or falsity, by means of its pictorial form.

2.221 What a picture represents is its sense.

2.222 The agreement or disagreement or its sense with reality constitutes its truth or falsity.

2.223 In order to tell whether a picture is true or false we must compare it with reality.

2.224 It is impossible to tell from the picture alone whether it is true or false.

2.225 There are no pictures that are true a priori.

3 A logical picture of facts is a thought.

3.001 'A state of affairs is thinkable': what this means is that we can picture it to ourselves.

3.01 The totality of true thoughts is a picture of the world.

3.02 A thought contains the possibility of the situation of which it is the thought. What is thinkable is possible too.

3.03 Thought can never be of anything illogical, since, if it were, we should have to think illogically.

3.031 It used to be said that God could create anything except what would be contrary to the laws of logic. The truth is that we could not say what an 'illogical' world would look like.

3.032 It is as impossible to represent in language anything that 'contradicts logic' as it is in geometry to represent by its coordinates a figure that contradicts the laws of space, or to give the coordinates of a point that does not exist.

3.0321 Though a state of affairs that would contravene the laws of physics can be represented by us spatially, one that would contravene the laws of geometry cannot.

3.04 It a thought were correct a priori, it would be a thought whose possibility ensured its truth.

3.05 A priori knowledge that a thought was true would be possible only it its truth were recognizable from the thought itself (without anything a to compare it with).

3.1 In a proposition a thought finds an expression that can be perceived by the senses.

3.11 We use the perceptible sign of a proposition (spoken or written, etc.) as a projection of a possible situation. The method of projection is to think of the sense of the proposition.

3.12 I call the sign with which we express a thought a propositional sign. And a proposition is a propositional sign in its projective relation to the world.

3.13 A proposition, therefore, does not actually contain its sense, but does contain the possibility of expressing it. ('The content of a proposition' means the content of a proposition that has sense.) A proposition contains the form, but not the content, of its sense.

3.14 What constitutes a propositional sign is that in its elements (the words) stand in a determinate relation to one another. A propositional sign is a fact.

3.141 A proposition is not a blend of words. (Just as a theme in music is not a blend of notes.) A proposition is articulate.

3.142 Only facts can express a sense, a set of names cannot.

3.143 Although a propositional sign is a fact, this is obscured by the usual form of expression in writing or print. For in a printed proposition, for example, no essential difference is apparent between a propositional sign and a word. (That is what made it possible for Frege to call a proposition a composite name.)

3.1431 The essence of a propositional sign is very clearly seen if we imagine one composed of spatial objects (such as tables, chairs, and books) instead of written signs.

3.1432 Instead of, 'The complex sign "aRb" says that a stands to b in the relation R' we ought to put, 'That "a" stands to "b" in a certain relation says that aRb.'

3.144 Situations can be described but not given names.

3.2 In a proposition a thought can be expressed in such a way that elements of the propositional sign correspond to the objects of the thought.

3.201 I call such elements 'simple signs', and such a proposition 'complete analysed'.

3.202 The simple signs employed in propositions are called names.

3.203 A name means an object. The object is its meaning. ('A' is the same sign as 'A'.)

3.21 The configuration of objects in a situation corresponds to the configuration of simple signs in the propositional sign.

3.221 Objects can only be named. Signs are their representatives. I can only speak about them: I cannot put them into words. Propositions can only say how things are, not what they are.

3.23 The requirement that simple signs be possible is the requirement that sense be determinate.

3.24 A proposition about a complex stands in an internal relation to a proposition about a constituent of the complex. A complex can be given only by its description, which will be right or wrong. A proposition that mentions a complex will not be nonsensical, if the complex does not exits, but simply false. When a propositional element signifies a complex, this can be seen from an indeterminateness in the propositions in which it occurs. In such cases we know that the proposition leaves something undetermined. (In fact the notation for generality contains a prototype.) The contraction of a symbol for a complex into a simple symbol can be expressed in a definition.

3.25 A proposition cannot be dissected any further by means of a definition: it is a primitive sign.

3.261 Every sign that has a definition signifies via the signs that serve to define it; and the definitions point the way. Two signs cannot signify in the same manner if one is primitive and the other is defined by means of primitive signs. Names cannot be anatomized by means of definitions. (Nor can any sign that has a meaning independently and on its own.)

3.262 What signs fail to express, their application shows. What signs slur over, their application says clearly.

3.263 The meanings of primitive signs can be explained by means of elucidations. Elucidations are propositions that stood if the meanings of those signs are already known.

3.3 Only propositions have sense; only in the nexus of a proposition does a name have meaning.

3.31 I call any part of a proposition that characterizes its sense an expression (or a symbol). (A proposition is itself an expression.) Everything essential to their sense that propositions can have in common with one another is an expression. An expression is the mark of a form and a content.

3.311 An expression presupposes the forms of all the propositions in which it can occur. It is the common characteristic mark of a class of propositions.

3.312 It is therefore presented by means of the general form of the propositions that it characterizes. In fact, in this form the expression will be constant and everything else variable.

3.313 Thus an expression is presented by means of a variable whose values are the propositions that contain the expression. (In the limiting case the variable becomes a constant, the expression becomes a proposition.) I call such a variable a 'propositional variable'.

3.314 An expression has meaning only in a proposition. All variables can be construed as propositional variables. (Even variable names.)

3.315 If we turn a constituent of a proposition into a variable, there is a class of propositions all of which are values of the resulting variable proposition. In general, this class too will be dependent on the meaning that our arbitrary conventions have given to parts of the original proposition. But if all the signs in it that have arbitrarily determined meanings are turned into variables, we shall still get a class of this kind. This one, however, is not dependent on any convention, but solely on the nature of the pro position. It corresponds to a logical form—a logical prototype.

3.316 What values a propositional variable may take is something that is stipulated. The stipulation of values is the variable.

3.317 To stipulate values for a propositional variable is to give the propositions whose common characteristic the variable is. The stipulation is a description of those propositions. The stipulation will therefore be concerned only with symbols, not with their meaning. And the only thing essential to the stipulation is that it is merely a description of symbols and states nothing about what is signified. How the description of the propositions is produced is not essential.

3.318 Like Frege and Russell I construe a proposition as a function of the expressions contained in it.

3.32 A sign is what can be perceived of a symbol.

3.321 So one and the same sign (written or spoken, etc.) can be common to two different symbols—in which case they will signify in different ways.

3.322 Our use of the same sign to signify two different objects can never indicate a common characteristic of the two, if we use it with two different modes of signification. For the sign, of course, is arbitrary. So we could choose two different signs instead, and then what would be left in common on the signifying side?

3.323 In everyday language it very frequently happens that the same word has different modes of signification—and so belongs to different symbols—or that two words that have different modes of signification are employed in propositions in what is superficially the same way. Thus the word 'is' figures as the copula, as a sign for identity, and as an expression for existence; 'exist' figures as an intransitive verb like 'go', and 'identical' as an adjective; we speak of something, but also of something's happening. (In the proposition, 'Green is green'—where the first word is the proper name of a person and the last an adjective—these words do not merely have different meanings: they are different symbols.)

3.324 In this way the most fundamental confusions are easily produced (the whole of philosophy is full of them).

3.325 In order to avoid such errors we must make use of a sign-language that excludes them by not using the same sign for different symbols and by not using in a superficially similar way signs that have different modes of signification: that is to say, a sign-language that is governed by logical grammar—by logical syntax. (The conceptual notation of Frege and Russell is such a language, though, it is true, it fails to exclude all mistakes.)

3.326 In order to recognize a symbol by its sign we must observe how it is used with a sense.

3.327 A sign does not determine a logical form unless it is taken together with its logico-syntactical employment.

3.328 If a sign is useless, it is meaningless. That is the point of Occam's maxim. (If everything behaves as if a sign had meaning, then it does have meaning.)

3.33 In logical syntax the meaning of a sign should never play a role. It must be possible to establish logical syntax without mentioning the meaning of a sign: only the description of expressions may be presupposed.

3.331 From this observation we turn to Russell's 'theory of types'. It can be seen that Russell must be wrong, because he had to mention the meaning of signs when establishing the rules for them.

3.332 No proposition can make a statement about itself, because a propositional sign cannot be contained in itself (that is the whole of the 'theory of types').

3.333 The reason why a function cannot be its own argument is that the sign for a function already contains the prototype of its argument, and it cannot contain itself. For let us suppose that the function $F(fx)$ could be its own argument: in that case there would be a proposition '$F(F(fx))$', in which the outer function F and the inner function F must have different meanings, since the inner one has the form $O(f(x))$ and the outer one has the form $Y(O(fx))$. Only the letter 'F' is common to the two functions, but the letter by itself signifies nothing. This immediately becomes clear if instead of '$F(Fu)$' we write '$(do) : F(Ou) . Ou = Fu$'. That disposes of Russell's paradox.

3.334 The rules of logical syntax must go without saying, once we know how each individual sign signifies.

3.34 A proposition possesses essential and accidental features. Accidental features are those that result from the particular way in which the propositional sign is produced. Essential features are those without which the proposition could not express its sense.

3.341 So what is essential in a proposition is what all propositions that can express the same sense have in common. And similarly, in general, what is essential in a symbol is what all symbols that can serve the same purpose have in common.

3.3411 So one could say that the real name of an object was what all symbols that signified it had in common. Thus, one by one, all kinds of composition would prove to be unessential to a name.

3.342 Although there is something arbitrary in our notations, this much is not arbitrary—that when we have determined one thing arbitrarily, something else is necessarily the case. (This derives from the essence of notation.)

3.3421 A particular mode of signifying may be unimportant but it is always important that it is a possible mode of signifying. And that is generally so in philosophy: again and again the individual case turns out to be unimportant, but the possibility of each individual case discloses something about the essence of the world.

3.343 Definitions are rules for translating from one language into another. Any correct sign-language must be translatable into any other in accordance with such rules: it is this that they all have in common.

3.344 What signifies in a symbol is what is common to all the symbols that the rules of logical syntax allow us to substitute for it.

3.3441 For instance, we can express what is common to all notations for truth-functions in the following way: they have in common that, for example, the notation that uses 'Pp' ('not p') and 'p C g' ('p or g') can be substituted for any of them. (This serves to characterize the way in which something general can be disclosed by the possibility of a specific notation.)

3.3442 Nor does analysis resolve the sign for a complex in an arbitrary way, so that it would have a different resolution every time that it was incorporated in a different proposition.

3.4 A proposition determines a place in logical space. The existence of this logical place is guaranteed by the mere existence of the constituents—by the existence of the proposition with a sense.

3.41 The propositional sign with logical co-ordinates—that is the logical place.

3.411 In geometry and logic alike a place is a possibility: something can exist in it.

3.42 A proposition can determine only one place in logical space: nevertheless the whole of logical space must already be given by it. (Otherwise negation, logical sum, logical product, etc., would introduce more and more new elements in co-ordination.) (The logical scaffolding surrounding a picture determines logical space. The force of a proposition reaches through the whole of logical space.)

3.5 A propositional sign, applied and thought out, is a thought.

4 A thought is a proposition with a sense.

4.001 The totality of propositions is language.

4.022 Man possesses the ability to construct languages capable of expressing every sense, without having any idea how each word has meaning or what its meaning is—just as people speak without knowing how the individual sounds are produced. Everyday language is a part of the human organism and is no less complicated than it. It is not humanly possible to gather immediately from it what the logic of language is. Language disguises thought. So much so, that from the outward form of the clothing it is impossible to infer the form of the thought beneath it, because the outward form of the clothing is not designed to reveal the form of the body, but for entirely different purposes.

The tacit conventions on which the understanding of everyday language depends are enormously complicated.

4.003 Most of the propositions and questions to be found in philosophical works are not false but nonsensical. Consequently we cannot give any answer to questions of this kind, but can only point out that they are nonsensical. Most of the propositions and questions of philosophers arise from our failure to understand the logic of our language. (They belong to the same class as the question whether the good is more or less identical than the beautiful.) And it is not surprising that the deepest problems are in fact not problems at all.

4.0031 All philosophy is a 'critique of language' (though not in Mauthner's sense). It was Russell who performed the service of showing that the apparent logical form of a proposition need not be its real one.

4.01 A proposition is a picture of reality. A proposition is a model of reality as we imagine it.

4.011 At first sight a proposition—one set out on the printed page, for example—does not seem to be a picture of the reality with which it is concerned. But neither do written notes seem at first sight to be a picture of a piece of music, nor our phonetic notation (the alphabet) to be a picture of our speech. And yet these sign-languages prove to be pictures, even in the ordinary sense, of what they represent.

4.012 It is obvious that a proposition of the form 'aRb' strikes us as a picture. In this case the sign is obviously a likeness of what is signified.

4.013 And if we penetrate to the essence of this pictorial character, we see that it is not impaired by apparent irregularities (such as the use [sharp] of and [flat] in musical notation). For even these irregularities depict what they are intended to express; only they do it in a different way.

4.014 A gramophone record, the musical idea, the written notes, and the sound-waves, all stand to one another in the same internal relation of depicting that holds between language and the world. They are all constructed according to a common logical pattern. (Like the two youths in the fairy-tale, their two horses, and their lilies. They are all in a certain sense one.)

4.0141 There is a general rule by means of which the musician can obtain the symphony from the score, and which makes it possible to derive the symphony from the groove on the gramophone record, and, using the first rule, to derive the score again. That is what constitutes the inner similarity between these things which seem to be constructed in such entirely different ways. And that rule is the law of projection which projects the symphony into the language of musical notation. It is the rule for translating this language into the language of gramophone records.

4.015 The possibility of all imagery, of all our pictorial modes of expression, is contained in the logic of depiction.

4.016 In order to understand the essential nature of a proposition, we should consider hieroglyphic script, which depicts the facts that it describes. And alphabetic script developed out of it without losing what was essential to depiction.

4.02 We can see this from the fact that we understand the sense of a propositional sign without its having been explained to us.

4.021 A proposition is a picture of reality: for if I understand a proposition, I know the situation that it represents. And I understand the proposition without having had its sense explained to me.

4.022 A proposition shows its sense. A proposition shows how things stand if it is true. And it says that they do so stand.

4.023 A proposition must restrict reality to two alternatives: yes or no. In order to do that, it must describe reality completely. A proposition is a description of a state of affairs. Just as a description of an object describes it by giving its external properties, so a proposition describes reality by its internal properties. A proposition constructs a world with the help of a logical scaffolding, so that one can actually see from the proposition how everything stands logically if it is true. One can draw inferences from a false proposition.

4.024 To understand a proposition means to know what is the case if it is true. (One can understand it, therefore, without knowing whether it is true.) It is understood by anyone who understands its constituents.

4.025 When translating one language into another, we do not proceed by translating each proposition of the one into a proposition of the other, but merely by translating the constituents of propositions. (And the dictionary translates not only substantives, but also verbs, adjectives, and conjunctions, etc.; and it treats them all in the same way.)

4.026 The meanings of simple signs (words) must be explained to us if we are to understand them. With propositions, however, we make ourselves understood.

4.027 It belongs to the essence of a proposition that it should be able to communicate a new sense to us.

4.03 A proposition must use old expressions to communicate a new sense. A proposition communicates a situation to us, and so it must be essentially connected with the situation. And the connexion is precisely that it is its logical picture. A proposition states something only in so far as it is a picture.

4.031 In a proposition a situation is, as it were, constructed by way of experiment. Instead of, 'This proposition has such and such a sense, we can simply say, 'This proposition represents such and such a situation'.

4.0311 One name stands for one thing, another for another thing, and they are combined with one another. In this way the whole group—like a tableau vivant—presents a state of affairs.

4.0312 The possibility of propositions is based on the principle that objects have signs as their representatives. My fundamental idea is that the 'logical constants' are not representatives; that there can be no representatives of the logic of facts.

4.032 It is only in so far as a proposition is logically articulated that it is a picture of a situation. (Even the proposition, 'Ambulo', is composite: for its stem with a different ending yields a different sense, and so does its ending with a different stem.)

4.04 In a proposition there must be exactly as many distinguishable parts as in the situation that it represents. The two must possess the same logical (mathematical) multiplicity. (Compare Hertz's Mechanics on dynamical models.)

4.041 This mathematical multiplicity, of course, cannot itself be the subject of depiction. One cannot get away from it when depicting.

4.0411 If, for example, we wanted to express what we now write as '(x) . fx' by putting an affix in front of 'fx'—for instance by writing 'Gen. fx'- -it would not be adequate: we should not know what was being generalized. If we wanted to signalize it with an affix 'g'—for instance by writing 'f(xg)'—that would not be adequate either: we should not know the scope of the generality-sign. If we were to try to do it by introducing a mark into the argument-places—for instance by writing '(G,G) . F(G,G)'—it would not be adequate: we should not be able to establish the identity of the variables. And so on. All these modes of signifying are inadequate because they lack the necessary mathematical multiplicity.

4.0412 For the same reason the idealist's appeal to 'spatial spectacles' is inadequate to explain the seeing of spatial relations, because it cannot explain the multiplicity of these relations.

4.05 Reality is compared with propositions.

4.06 A proposition can be true or false only in virtue of being a picture of reality.

4.061 It must not be overlooked that a proposition has a sense that is independent of the facts: otherwise one can easily suppose that true and false are relations of equal status between signs and what they signify. In that case one could say, for example, that 'p' signified in the true way what 'Pp' signified in the false way, etc.

4.062 Can we not make ourselves understood with false propositions just as we have done up till now with true ones?—So long as it is known that they are meant to be false.—No! For a proposition is true if we use it to say that things stand in a certain way, and they do; and if by 'p' we mean Pp and things stand as we mean that they do, then, construed in the new way, 'p' is true and not false.

4.0621 But it is important that the signs 'p' and 'Pp' can say the same thing. For it shows that nothing in reality corresponds to the sign 'P'. The occurrence of negation in a proposition is not enough to characterize its sense (PPp = p). The propositions 'p' and 'Pp' have opposite sense, but there corresponds to them one and the same reality.

4.063 An analogy to illustrate the concept of truth: imagine a black spot on white paper: you can describe the shape of the spot by saying, for each point on the sheet, whether it is black or white. To the fact that a point is black there corresponds a positive fact, and to the fact that a point is white (not black), a negative fact. If I designate a point on the sheet (a truth-value according to Frege), then this corresponds to the supposition that is put forward for judgement, etc. etc. But in order to be able to say that a point is black or white, I must first know when a point is called black, and when white: in order to be able to say,'"p" is true (or false)', I must have determined in what circumstances I call 'p' true, and in so doing I determine the sense of the proposition. Now the point where the simile breaks down is this: we can indicate a point on the paper even if we do not know what black and white are, but if a proposition has no sense, nothing corresponds to it, since it does not designatea thing (a) which might have properties called 'false' or 'true'. The verb of a proposition is not 'is true' or 'is false', as Frege thought: rather, that which 'is true' must already contain the verb.

4.064 Every proposition must already have a sense: it cannot be given a sense by affirmation. Indeed its sense is just what is affirmed. And the same applies to negation, etc.

4.0641 One could say that negation must be related to the logical place determined by the negated proposition. The negating proposition determines a logical place different from that of the negated proposition. The negating proposition determines a logical place with the help of the logical place of the negated proposition. For it describes it as lying outside the latter's logical place. The negated proposition can be negated again, and this in itself shows that what is negated is already a proposition, and not merely something that is prelimary to a proposition.

4.1 Propositions represent the existence and non-existence of states of affairs.

4.11 The totality of true propositions is the whole of natural science (or the whole corpus of the natural sciences).

4.111 Philosophy is not one of the natural sciences. (The word 'philosophy' must mean something whose place is above or below the natural sciences, not beside them.)

4.112 Philosophy aims at the logical clarification of thoughts. Philosophy is not a body of doctrine but an activity. A philosophical work consists essentially of elucidations. Philosophy does not result in 'philosophical propositions', but rather in the clarification of propositions. Without philosophy thoughts are, as it were, cloudy and indistinct: its task is to make them clear and to give them sharp boundaries.

4.1121 Psychology is no more closely related to philosophy than any other natural science. Theory of knowledge is the philosophy of psychology. Does not my study of sign-language correspond to the study of thought-processes, which philosophers used to consider so essential to the philosophy of logic? Only in most cases they got entangled in unessential psychological investigations, and with my method too there is an analogous risk.

4.1122 Darwin's theory has no more to do with philosophy than any other hypothesis in natural science.

4.113 Philosophy sets limits to the much disputed sphere of natural science.

4.114 It must set limits to what can be thought; and, in doing so, to what cannot be thought. It must set limits to what cannot be thought by working outwards through what can be thought.

4.115 It will signify what cannot be said, by presenting clearly what can be said.

4.116 Everything that can be thought at all can be thought clearly. Everything that can be put into words can be put clearly.

4.12 Propositions can represent the whole of reality, but they cannot represent what they must have in common with reality in order to be able to represent it—logical form. In order to be able to represent logical form, we should have to be able to station ourselves with propositions somewhere outside logic, that is to say outside the world.

4.121 Propositions cannot represent logical form: it is mirrored in them. What finds its reflection in language, language cannot represent. What expresses itself in language, we cannot express by means of language. Propositions show the logical form of reality. They display it.

4.1211 Thus one proposition 'fa' shows that the object a occurs in its sense, two propositions 'fa' and 'ga' show that the same object is mentioned in both of them. If two propositions contradict one another, then their structure shows it; the same is true if one of them follows from the other. And so on.

4.1212 What can be shown, cannot be said.

4.1213 Now, too, we understand our feeling that once we have a in which everything is all right, we already have a correct logical point of view.

4.122 In a certain sense we can talk about formal properties of objects and states of affairs, or, in the case of facts, about structural properties: and in the same sense about formal relations and structural relations. (Instead of 'structural property' I also say 'internal property'; instead of 'structural relation', 'internal relation'. I introduce these expressions in order to indicate the source of the confusion between internal relations and relations proper (external relations), which is very widespread among philosophers.) It is impossible, however, to assert by means of propositions that such internal properties and relations obtain: rather, this makes itself manifest in the propositions that represent the relevant states of affairs and are concerned with the relevant objects.

4.1221 An internal property of a fact can also be bed a feature of that fact (in the sense in which we speak of facial features, for example).

4.123 A property is internal if it is unthinkable that its object should not possess it. (This shade of blue and that one stand, eo ipso, in the internal relation of lighter to darker. It is unthinkable that these two objects should not stand in this relation.) (Here the shifting use of the word 'object' corresponds to the shifting use of the words 'property' and 'relation'.)

4.124 The existence of an internal property of a possible situation is not expressed by means of a proposition: rather, it expresses itself in the proposition representing the situation, by means of an internal property of that proposition. It would be just as nonsensical to assert that a proposition had a formal property as to deny it.

4.1241 It is impossible to distinguish forms from one another by saying that one has this property and another that property: for this presupposes that it makes sense to ascribe either property to either form.

4.125 The existence of an internal relation between possible situations expresses itself in language by means of an internal relation between the propositions representing them.

4.1251 Here we have the answer to the vexed question 'whether all relations are internal or external'.

4.1252 I call a series that is ordered by an internal relation a series of forms. The order of the number-series is not governed by an external relation but by an internal relation. The same is true of the series of propositions 'aRb', '(d : c) : aRx . xRb', '(d x,y) : aRx . xRy . yRb', and so forth. (If b stands in one of these relations to a, I call b a successor of a.)

4.126 We can now talk about formal concepts, in the same sense that we speak of formal properties. (I introduce this expression in order to exhibit the source of the confusion between formal concepts and concepts proper, which pervades the whole of traditional logic.) When something falls under a formal concept as one of its objects, this cannot be expressed by means of a proposition. Instead it is shown in the very sign for this object. (A name shows that it signifies an object, a sign for a number that it signifies a number, etc.) Formal concepts cannot, in fact, be represented by means of a function, as concepts proper can. For their characteristics, formal properties, are not expressed by means of functions. The expression for a formal property is a feature of certain symbols. So the sign for the characteristics of a formal concept is a distinctive feature of all symbols whose meanings fall under the concept. So the expression for a formal concept is a propositional variable in which this distinctive feature alone is constant.

4.127 The propositional variable signifies the formal concept, and its values signify the objects that fall under the concept.

4.1271 Every variable is the sign for a formal concept. For every variable represents a constant form that all its values possess, and this can be regarded as a formal property of those values.

4.1272 Thus the variable name 'x' is the proper sign for the pseudo-concept object. Wherever the word 'object' ('thing', etc.) is correctly used, it is expressed in conceptual notation by a variable name. For example, in the proposition, 'There are 2 objects which. . .', it is expressed by ' (dx,y) ... '. Wherever it is used in a different way, that is as a proper , nonsensical pseudo-propositions are the result. So one cannot say, for example, 'There are objects', as one might say, 'There are books'. And it is just as impossible to say, 'There are 100 objects', or, 'There are !0 objects'. And it is nonsensical to speak of the total number of objects. The same applies to the words 'complex', 'fact', 'function', 'number', etc. They all signify formal concepts, and are represented in conceptual notation by variables, not by functions or classes (as Frege and Russell believed). '1 is a number', 'There is only one zero', and all similar expressions are nonsensical. (It is just as nonsensical to say, 'There is only one 1', as it would be to say, '2 + 2 at 3 o'clock equals 4'.)

4.12721 A formal concept is given immediately any object falling under it is given. It is not possible, therefore, to introduce as primitive ideas objects belonging to a formal concept and the formal concept itself. So it is impossible, for example, to introduce as primitive ideas both the concept of a function and specific functions, as Russell does; or the concept of a number and particular numbers.

4.1273 If we want to express in conceptual notation the general proposition, 'b is a successor of a', then we require an expression for the general term of the series of forms 'aRb', '(d : c) : aRx . xRb', '(d x,y) : aRx . xRy . yRb', ... , In order to express the general term of a series of forms, we must use a variable, because the concept 'term of that series of forms' is a formal concept. (This is what Frege and Russell overlooked: consequently the way in which they want to express general propositions like the one above is incorrect; it contains a vicious circle.) We can determine the general term of a series of forms by giving its first term and the general form of the operation that produces the next term out of the proposition that precedes it.

4.1274 To ask whether a formal concept exists is nonsensical. For no proposition can be the answer to such a question. (So, for example, the question, 'Are there unanalysable subject-predicate propositions?' cannot be asked.)

4.128 Logical forms are without number. Hence there are no preeminent numbers in logic, and hence there is no possibility of philosophical monism or dualism, etc.

4.2 The sense of a proposition is its agreement and disagreement with possibilities of existence and non-existence of states of affairs. 4.21 The simplest kind of proposition, an elementary proposition, asserts the existence of a state of affairs.

4.211 It is a sign of a proposition's being elementary that there can be no elementary proposition contradicting it.

4.22 An elementary proposition consists of names. It is a nexus, a concatenation, of names.

4.221 It is obvious that the analysis of propositions must bring us to elementary propositions which consist of names in immediate combination. This raises the question how such combination into propositions comes about.

4.2211 Even if the world is infinitely complex, so that every fact consists of infinitely many states of affairs and every state of affairs is composed of infinitely many objects, there would still have to be objects and states of affairs.

4.23 It is only in the nexus of an elementary proposition that a name occurs in a proposition.

4.24 Names are the simple symbols: I indicate them by single letters ('x', 'y', 'z'). I write elementary propositions as functions of names, so that they have the form 'fx', 'O (x,y)', etc. Or I indicate them by the letters 'p', 'q', 'r'.

4.241 When I use two signs with one and the same meaning, I express this by putting the sign '=' between them. So 'a = b' means that the sign 'b' can be substituted for the sign 'a'. (If I use an equation to introduce a new sign 'b', laying down that it shall serve as a substitute for a sign a that is already known, then, like Russell, I write the equation—definition—in the form 'a = b Def.' A definition is a rule dealing with signs.)

4.242 Expressions of the form 'a = b' are, therefore, mere representational devices. They state nothing about the meaning of the signs 'a' and 'b'.

4.243 Can we understand two names without knowing whether they signify the same thing or two different things?—Can we understand a proposition in which two names occur without knowing whether their meaning is the same or different? Suppose I know the meaning of an English word and of a German word that means the same: then it is impossible for me to be unaware that they do mean the same; I must be capable of translating each into the other. Expressions like 'a = a', and those derived from them, are neither elementary propositions nor is there any other way in which they have sense. (This will become evident later.)

4.25 If an elementary proposition is true, the state of affairs exists: if an elementary proposition is false, the state of affairs does not exist.

4.26 If all true elementary propositions are given, the result is a complete description of the world. The world is completely described by giving all elementary propositions, and adding which of them are true and which false. For n states of affairs, there are possibilities of existence and non-existence. Of these states of affairs any combination can exist and the remainder not exist.

4.28 There correspond to these combinations the same number of possibilities of truth—and falsity—for n elementary propositions.

4.3 Truth-possibilities of elementary propositions mean Possibilities of existence and non-existence of states of affairs.

4.31 We can represent truth-possibilities by schemata of the following kind ('T' means 'true', 'F' means 'false'; the rows of 'T's and 'F's under the row of elementary propositions symbolize their truth-possibilities in a way that can easily be understood):

4.4 A proposition is an expression of agreement and disagreement with of elementary propositions.

4.41 Truth-possibilities of elementary propositions are the conditions of the truth and falsity of propositions.

4.411 It immediately strikes one as probable that the introduction of elementary propositions provides the basis for understanding all other kinds of proposition. Indeed the understanding of general propositions palpably depends on the understanding of elementary propositions.

4.42 For n elementary propositions there are ways in which a proposition can agree and disagree with their truth possibilities.

4.43 We can express agreement with truth-possibilities by correlating the mark 'T' (true) with them in the schema. The absence of this mark means disagreement.

4.431 The expression of agreement and disagreement with the truth possibilities of elementary propositions expresses the truth-conditions of a proposition. A proposition is the expression of its truth-conditions. (Thus Frege was quite right to use them as a starting point when he explained the signs of his conceptual notation. But the explanation of the concept of truth that Frege gives is mistaken: if 'the true' and 'the false' were really objects, and were the arguments in Pp etc., then Frege's method of determining the sense of 'Pp' would leave it absolutely undetermined.)

4.44 The sign that results from correlating the mark 'I" with is a propositional sign.

4.441 It is clear that a complex of the signs 'F' and 'T' has no object (or complex of objects) corresponding to it, just as there is none corresponding to the horizontal and vertical lines or to the brackets.—There are no 'logical objects'. Of course the same applies to all signs that express what the schemata of 'T's' and 'F's' express.

4.442 For example, the following is a propositional sign: (Frege's 'judgement stroke' '|-' is logically quite meaningless: in the works of Frege (and Russell) it simply indicates that these authors hold the propositions marked with this sign to be true. Thus '|-' is no more a component part of a proposition than is, for instance, the proposition's number. It is quite impossible for a proposition to state that it itself is true.) If the order or the truth-possibilities in a scheme is fixed once and for all by a combinatory rule, then the last column by itself will be an expression of the truth-conditions. If we now write this column as a row, the propositional sign will become '(TT-T) (p,q)' or more explicitly '(TTFT) (p,q)' (The number of places in the left-hand pair of brackets is determined by the number of terms in the right-hand pair.)

4.45 For n elementary propositions there are Ln possible groups of . The groups of truth-conditions that are obtainable from the truth-possibilities of a given number of elementary propositions can be arranged in a series.

4.46 Among the possible groups of truth-conditions there are two extreme cases. In one of these cases the proposition is true for all the of the elementary propositions. We say that the are tautological. In the second case the proposition is false for all the truth-possibilities: the truth-conditions are contradictory . In the first case we call the proposition a tautology; in the second, a contradiction.

4.461 Propositions show what they say; tautologies and contradictions show that they say nothing. A tautology has no truth-conditions, since it is unconditionally true: and a contradiction is true on no condition. Tautologies and contradictions lack sense. (Like a point from which two arrows go out in opposite directions to one another.) (For example, I know nothing about the weather when I know that it is either raining or not raining.)

4.46211 Tautologies and contradictions are not, however, nonsensical. They are part of the symbolism, much as '0' is part of the symbolism of arithmetic.

4.462 Tautologies and contradictions are not pictures of reality. They do not represent any possible situations. For the former admit all possible situations, and latter none . In a tautology the conditions of agreement with the world—the representational relations—cancel one another, so that it does not stand in any representational relation to reality.

4.463 The truth-conditions of a proposition determine the range that it leaves open to the facts. (A proposition, a picture, or a model is, in the negative sense, like a solid body that restricts the freedom of movement of others, and in the positive sense, like a space bounded by solid substance in which there is room for a body.) A tautology leaves open to reality the whole—the infinite whole—of logical space: a contradiction fills the whole of logical space leaving no point of it for reality. Thus neither of them can determine reality in any way.

4.464 A tautology's truth is certain, a proposition's possible, a contradiction's impossible. (Certain, possible, impossible: here we have the first indication of the scale that we need in the theory of probability.)

4.465 The logical product of a tautology and a proposition says the same thing as the proposition. This product, therefore, is identical with the proposition. For it is impossible to alter what is essential to a symbol without altering its sense.

4.466 What corresponds to a determinate logical combination of signs is a determinate logical combination of their meanings. It is only to the uncombined signs that absolutely any combination corresponds. In other words, propositions that are true for every situation cannot be combinations of signs at all, since, if they were, only determinate combinations of objects could correspond to them. (And what is not a logical combination has no combination of objects corresponding to it.) Tautology and contradiction are the limiting cases—indeed the disintegration—of the combination of signs.

4.4661 Admittedly the signs are still combined with one another even in tautologies and contradictions—i.e. they stand in certain relations to one another: but these relations have no meaning, they are not essential to the symbol .

4.5 It now seems possible to give the most general propositional form: that is, to give a description of the propositions of any sign-language whatsoever in such a way that every possible sense can be expressed by a symbol satisfying the description, and every symbol satisfying the description can express a sense, provided that the meanings of the names are suitably chosen. It is clear that only what is essential to the most general propositional form may be included in its description—for otherwise it would not be the most general form. The existence of a general propositional form is proved by the fact that there cannot be a proposition whose form could not have been foreseen (i.e. constructed). The general form of a proposition is: This is how things stand.

4.51 Suppose that I am given all elementary propositions: then I can simply ask what propositions I can construct out of them. And there I have all propositions, and that fixes their limits.

4.52 Propositions comprise all that follows from the totality of all elementary propositions (and, of course, from its being the totality of them all). (Thus, in a certain sense, it could be said that all propositions were generalizations of elementary propositions.)

4.53 The general propositional form is a variable.

5 A proposition is a truth-function of elementary propositions. (An elementary proposition is a truth-function of itself.)

5.01 Elementary propositions are the truth-arguments of propositions.

5.02 The arguments of functions are readily confused with the affixes of names. For both arguments and affixes enable me to recognize the meaning of the signs containing them. For example, when Russell writes '+c', the 'c' is an affix which indicates that the sign as a whole is the addition-sign for cardinal numbers. But the use of this sign is the result of arbitrary convention and it would be quite possible to choose a simple sign instead of '+c'; in 'Pp' however, 'p' is not an affix but an argument: the sense of 'Pp' cannot be understood unless the sense of 'p' has been understood already. (In the name Julius Caesar 'Julius' is an affix. An affix is always part of a description of the object to whose name we attach it: e.g. the Caesar of the Julian gens.) If I am not mistaken, Frege's theory about the meaning of propositions and functions is based on the confusion between an argument and an affix. Frege regarded the propositions of logic as names, and their arguments as the affixes of those names.

5.1 Truth-functions can be arranged in series. That is the foundation of the theory of probability.

5.101 The truth-functions of a given number of elementary propositions can always be set out in a schema of the following kind: (TTTT) (p, q) Tautology (If p then p, and if q then q.) (p z p . q z q) (FTTT) (p, q) In words : Not both p and q. (P(p . q)) (TFTT) (p, q) " : If q then p. (q z p) (TTFT) (p, q) " : If p then q. (p z q) (TTTF) (p, q) " : p or q. (p C q) (FFTT) (p, q) " : Not g. (Pq) (FTFT) (p, q) " : Not p. (Pp) (FTTF) (p, q) " : p or q, but not both. (p . Pq : C : q . Pp) (TFFT) (p, q) " : If p then p, and if q then p. (p + q) (TFTF) (p, q) " : p (TTFF) (p, q) " : q (FFFT) (p, q) " : Neither p nor q. (Pp . Pq or p | q) (FFTF) (p, q) " : p and not q. (p . Pq) (FTFF) (p, q) " : q and not p. (q . Pp) (TFFF) (p,q) " : q and p. (q . p) (FFFF) (p, q) Contradiction (p and not p, and q and not q.) (p . Pp . q . Pq) I will give the name truth-grounds of a proposition to those truth-possibilities of its truth-arguments that make it true.

5.11 If all the truth-grounds that are common to a number of propositions are at the same time truth-grounds of a certain proposition, then we say that the truth of that proposition follows from the truth of the others.

5.12 In particular, the truth of a proposition 'p' follows from the truth of another proposition 'q' is all the truth-grounds of the latter are of the former.

5.121 The truth-grounds of the one are contained in those of the other: p follows from q.

5.122 If p follows from q, the sense of 'p' is contained in the sense of 'q'.

5.123 If a god creates a world in which certain propositions are true, then by that very act he also creates a world in which all the propositions that follow from them come true. And similarly he could not create a world in which the proposition 'p' was true without creating all its objects.

5.124 A proposition affirms every proposition that follows from it.

5.1241 'p . q' is one of the propositions that affirm 'p' and at the same time one of the propositions that affirm 'q'. Two propositions are opposed to one another if there is no proposition with a sense, that affirms them both. Every proposition that contradicts another negate it.

5.13 When the truth of one proposition follows from the truth of others, we can see this from the structure of the proposition.

5.131 If the truth of one proposition follows from the truth of others, this finds expression in relations in which the forms of the propositions stand to one another: nor is it necessary for us to set up these relations between them, by combining them with one another in a single proposition; on the contrary, the relations are internal, and their existence is an immediate result of the existence of the propositions.

5.1311 When we infer q from p C q and Pp, the relation between the propositional forms of 'p C q' and 'Pp' is masked, in this case, by our mode of signifying. But if instead of 'p C q' we write, for example, 'p|q . | . p|q', and instead of 'Pp', 'p|p' (p|q = neither p nor q), then the inner connexion becomes obvious. (The possibility of inference from (x) . fx to fa shows that the symbol (x) . fx itself has generality in it.)

5.132 If p follows from q, I can make an inference from q to p, deduce p from q. The nature of the inference can be gathered only from the two propositions. They themselves are the only possible justification of the inference. 'Laws of inference', which are supposed to justify inferences, as in the works of Frege and Russell, have no sense, and would be superfluous.

5.133 All deductions are made a priori.

5.134 One elementary proposition cannot be deduced form another.

5.135 There is no possible way of making an inference form the existence of one situation to the existence of another, entirely different situation.

5.136 There is no causal nexus to justify such an inference.

5.1361 We cannot infer the events of the future from those of the present. Belief in the causal nexus is superstition.

5.1362 The freedom of the will consists in the impossibility of knowing actions that still lie in the future. We could know them only if causality were an inner necessity like that of logical inference.—The connexion between knowledge and what is known is that of logical necessity. ('A knows that p is the case', has no sense if p is a tautology.)

5.1363 If the truth of a proposition does not follow from the fact that it is self-evident to us, then its self-evidence in no way justifies our belief in its truth.

5.14 If one proposition follows from another, then the latter says more than the former, and the former less than the latter.

5.141 If p follows from q and q from p, then they are one and same proposition.

5.142 A tautology follows from all propositions: it says nothing.

5.143 Contradiction is that common factor of propositions which no proposition has in common with another. Tautology is the common factor of all propositions that have nothing in common with one another. Contradiction, one might say, vanishes outside all propositions: tautology vanishes inside them. Contradiction is the outer limit of propositions: tautology is the unsubstantial point at their centre.

5.15 If Tr is the number of the truth-grounds of a proposition 'r', and if Trs is the number of the truth-grounds of a proposition 's' that are at the same time truth-grounds of 'r', then we call the ratio Trs : Tr the degree of probability that the proposition 'r' gives to the proposition 's'. 5.151 In a schema like the one above in

5.101, let Tr be the number of 'T's' in the proposition r, and let Trs, be the number of 'T's' in the proposition s that stand in columns in which the proposition r has 'T's'. Then the proposition r gives to the proposition s the probability Trs : Tr.

5.1511 There is no special object peculiar to probability propositions.

5.152 When propositions have no truth-arguments in common with one another, we call them independent of one another. Two elementary propositions give one another the probability 1/2. If p follows from q, then the proposition 'q' gives to the proposition 'p' the probability 1. The certainty of logical inference is a limiting case of probability. (Application of this to tautology and contradiction.)

5.153 In itself, a proposition is neither probable nor improbable. Either an event occurs or it does not: there is no middle way.

5.154 Suppose that an urn contains black and white balls in equal numbers (and none of any other kind). I draw one ball after another, putting them back into the urn. By this experiment I can establish that the number of black balls drawn and the number of white balls drawn approximate to one another as the draw continues. So this is not a mathematical truth. Now, if I say, 'The probability of my drawing a white ball is equal to the probability of my drawing a black one', this means that all the circumstances that I know of (including the laws of nature assumed as hypotheses) give no more probability to the occurrence of the one event than to that of the other. That is to say, they give each the probability 1/2 as can easily be gathered from the above definitions. What I confirm by the experiment is that the occurrence of the two events is independent of the circumstances of which I have no more detailed knowledge.

5.155 The minimal unit for a probability proposition is this: The circumstances—of which I have no further knowledge—give such and such a degree of probability to the occurrence of a particular event.

5.156 It is in this way that probability is a generalization. It involves a general description of a propositional form. We use probability only in default of certainty—if our knowledge of a fact is not indeed complete, but we do know something about its form. (A proposition may well be an incomplete picture of a certain situation, but it is always a complete picture of something .) A probability proposition is a sort of excerpt from other propositions.

5.2 The structures of propositions stand in internal relations to one another.

5.21 In order to give prominence to these internal relations we can adopt the following mode of expression: we can represent a proposition as the result of an operation that produces it out of other propositions (which are the bases of the operation).

5.22 An operation is the expression of a relation between the structures of its result and of its bases.

5.23 The operation is what has to be done to the one proposition in order to make the other out of it.

5.231 And that will, of course, depend on their formal properties, on the internal similarity of their forms.

5.232 The internal relation by which a series is ordered is equivalent to the operation that produces one term from another.

5.233 Operations cannot make their appearance before the point at which one proposition is generated out of another in a logically meaningful way; i.e. the point at which the logical construction of propositions begins.

5.234 Truth-functions of elementary propositions are results of operations with elementary propositions as bases. (These operations I call .)

5.2341 The sense of a truth-function of p is a function of the sense of p. Negation, logical addition, logical multiplication, etc. etc. are operations. (Negation reverses the sense of a proposition.)

5.24 An operation manifests itself in a variable; it shows how we can get from one form of proposition to another. It gives expression to the difference between the forms. (And what the bases of an operation and its result have in common is just the bases themselves.)

5.241 An operation is not the mark of a form, but only of a difference between forms.

5.242 The operation that produces 'q' from 'p' also produces 'r' from 'q', and so on. There is only one way of expressing this: 'p', 'q', 'r', etc. have to be variables that give expression in a general way to certain formal relations.

5.25 The occurrence of an operation does not characterize the sense of a proposition. Indeed, no statement is made by an operation, but only by its result, and this depends on the bases of the operation. (Operations and functions must not be confused with each other.)

5.251 A function cannot be its own argument, whereas an operation can take one of its own results as its base.

5.252 It is only in this way that the step from one term of a series of forms to another is possible (from one type to another in the hierarchies of Russell and Whitehead). (Russell and Whitehead did not admit the possibility of such steps, but repeatedly availed themselves of it.)

5.2521 If an operation is applied repeatedly to its own results, I speak of successive applications of it. ('O'O'O'a' is the result of three successive applications of the operation 'O'E' to 'a'.) In a similar sense I speak of successive applications of more than one operation to a number of propositions.

5.2522 Accordingly I use the sign '[a, x, O'x]' for the general term of the series of forms a, O'a, O'O'a, ... This bracketed expression is a variable: the first term of the bracketed expression is the beginning of the series of forms, the second is the form of a term x arbitrarily selected from the series, and the third is the form of the term that immediately follows x in the series.

5.2523 The concept of successive applications of an operation is equivalent to the concept 'and so on'.

5.253 One operation can counteract the effect of another. Operations can cancel one another.

5.254 An operation can vanish (e.g. negation in 'PPp' : PPp = p).

5.3 All propositions are results of truth-operations on elementary propositions. A truth-operation is the way in which a truth-function is produced out of elementary propositions. It is of the essence of that, just as elementary propositions yield a truth-function of themselves, so too in the same way truth-functions yield a further. When a truth-operation is applied to truth-functions of elementary propositions, it always generates another truth-function of elementary propositions, another proposition. When a truth-operation is applied to the results of truth-operations on elementary propositions, there is always a single operation on elementary propositions that has the same result. Every proposition is the result of truth-operations on elementary propositions.

5.31 The schemata in 4.31 have a meaning even when 'p', 'q', 'r', etc. are not elementary propositions. And it is easy to see that the propositional sign in 4.442 expresses a single truth-function of elementary propositions even when 'p' and 'q' are truth-functions of elementary propositions.

5.32 All truth-functions are results of successive applications to elementary propositions of a finite number of truth-operations.

5.4 At this point it becomes manifest that there are no 'logical objects' or 'logical constants' (in Frege's and Russell's sense).

5.41 The reason is that the results of truth-operations on truth-functions are always identical whenever they are one and the same truth-function of elementary propositions.

5.42 It is self-evident that C, z, etc. are not relations in the sense in which right and left etc. are relations. The interdefinability of Frege's and Russell's 'primitive signs' of logic is enough to show that they are not primitive signs, still less signs for relations. And it is obvious that the 'z' defined by means of 'P' and 'C' is identical with the one that figures with 'P' in the definition of 'C'; and that the second 'C' is identical with the first one; and so on.

5.43 Even at first sight it seems scarcely credible that there should follow from one fact p infinitely many others, namely PPp,

PPPPp, etc. And it is no less remarkable that the infinite number of propositions of logic (mathematics) follow from half a dozen 'primitive propositions'. But in fact all the propositions of logic say the same thing, to wit nothing.

5.44 Truth-functions are not material functions. For example, an affirmation can be produced by double negation: in such a case does it follow that in some sense negation is contained in affirmation? Does 'PPp' negate Pp, or does it affirm p—or both? The proposition 'PPp' is not about negation, as if negation were an object: on the other hand, the possibility of negation is already written into affirmation. And if there were an object called 'P', it would follow that 'PPp' said something different from what 'p' said, just because the one proposition would then be about P and the other would not.

5.441 This vanishing of the apparent logical constants also occurs in the case of 'P(dx) . Pfx', which says the same as '(x) . fx', and in the case of '(dx) . fx . x = a', which says the same as 'fa'.

5.442 If we are given a proposition, then with it we are also given the results of all truth-operations that have it as their base.

5.45 If there are primitive logical signs, then any logic that fails to show clearly how they are placed relatively to one another and to justify their existence will be incorrect. The construction of logic out of its primitive signs must be made clear.

5.451 If logic has primitive ideas, they must be independent of one another. If a primitive idea has been introduced, it must have been introduced in all the combinations in which it ever occurs. It cannot, therefore, be introduced first for one combination and later reintroduced for another. For example, once negation has been introduced, we must understand it both in propositions of the form 'Pp' and in propositions like 'P(p C q)', '(dx) . Pfx', etc. We must not introduce it first for the one class of cases and then for the other, since it would then be left in doubt whether its meaning were the same in both cases, and no reason would have been given for combining the signs in the same way in both cases. (In short, Frege's remarks about introducing signs by means of definitions (in The Fundamental Laws of Arithmetic) also apply, mutatis mutandis, to the introduction of primitive signs.)

5.452 The introduction of any new device into the symbolism of logic is necessarily a momentous event. In logic a new device should not be introduced in brackets or in a footnote with what one might call a completely innocent air. (Thus in Russell and Whitehead's Principia Mathematica there occur definitions and primitive propositions expressed in words. Why this sudden appearance of words? It would require a justification, but none is given, or could be given, since the procedure is in fact illicit.) But if the introduction of a new device has proved necessary at a certain point, we must immediately ask ourselves, 'At what points is the employment of this device now unavoidable ?' and its place in logic must be made clear.

5.453 All numbers in logic stand in need of justification. Or rather, it must become evident that there are no numbers in logic. There are no numbers.

5.454 In logic there is no co-ordinate status, and there can be no classification. In logic there can be no distinction between the general and the specific.

5.4541 The solutions of the problems of logic must be simple, since they set the standard of simplicity. Men have always had a presentiment that there must be a realm in which the answers to questions are symmetrically combined—a priori—to form a self-contained system. A realm subject to the law: Simplex sigillum veri.

5.46 If we introduced logical signs properly, then we should also have introduced at the same time the sense of all combinations of them; i.e. not only 'p C q' but 'P(p C q)' as well, etc. etc. We should also have introduced at the same time the effect of all possible combinations of brackets. And thus it would have been made clear that the real general primitive signs are not ' p C q', '(dx) . fx', etc. but the most general form of their combinations.

5.461 Though it seems unimportant, it is in fact significant that the pseudo-relations of logic, such as C and z, need brackets—unlike real relations. Indeed, the use of brackets with these apparently primitive signs is itself an indication that they are not primitive signs. And surely no one is going to believe brackets have an independent meaning. 5.4611 Signs for logical operations are punctuation-marks,

5.47 It is clear that whatever we can say in advance about the form of all propositions, we must be able to say all at once . An elementary proposition really contains all logical operations in itself. For 'fa' says the same thing as '(dx) . fx . x = a' Wherever there is compositeness, argument and function are present, and where these are present, we already have all the logical constants. One could say that the sole logical constant was what all propositions, by their very nature, had in common with one another. But that is the general propositional form.

5.471 The general propositional form is the essence of a proposition.

5.4711 To give the essence of a proposition means to give the essence of all description, and thus the essence of the world.

5.472 The description of the most general propositional form is the description of the one and only general primitive sign in logic.

5.473 Logic must look after itself. If a sign is possible , then it is also capable of signifying. Whatever is possible in logic is also permitted. (The reason why 'Socrates is identical' means nothing is that there is no property called 'identical'. The proposition is nonsensical because we have failed to make an arbitrary determination, and not because the symbol, in itself, would be illegitimate.) In a certain sense, we cannot make mistakes in logic.

5.4731 Self-evidence, which Russell talked about so much, can become dispensable in logic, only because language itself prevents every logical mistake.—What makes logic a priori is the impossibility of illogical thought.

5.4732 We cannot give a sign the wrong sense.

5,47321 Occam's maxim is, of course, not an arbitrary rule, nor one that is justified by its success in practice: its point is that unnecessary units in a sign-language mean nothing. Signs that serve one purpose are logically equivalent, and signs that serve none are logically meaningless.

5.4733 Frege says that any legitimately constructed proposition must have a sense. And I say that any possible proposition is legitimately constructed, and, if it has no sense, that can only be because we have failed to give a meaning to some of its constituents. (Even if we think that we have done so.) Thus the reason why 'Socrates is identical' says nothing is that we have not given any adjectival meaning to the word 'identical'. For when it appears as a sign for identity, it symbolizes in an entirely different way—the signifying relation is a different one—therefore the symbols also are entirely different in the two cases: the two symbols have only the sign in common, and that is an accident.

5.474 The number of fundamental operations that are necessary depends solely on our notation.

5.475 All that is required is that we should construct a system of signs with a particular number of dimensions—with a particular mathematical multiplicity

5.476 It is clear that this is not a question of a number of primitive ideas that have to be signified, but rather of the expression of a rule.

5.5 Every truth-function is a result of successive applications to elementary propositions of the operation '(———-T)(E, ...)'. This operation negates all the propositions in the right-hand pair of brackets, and I call it the negation of those propositions.

5.501 When a bracketed expression has propositions as its terms—and the order of the terms inside the brackets is indifferent—then I indicate it by a sign of the form '(E)'. '(E)' is a variable whose values are terms of the bracketed expression and the bar over the variable indicates that it is the representative of ali its values in the brackets. (E.g. if E has the three values P,Q, R, then (E) = (P, Q, R).) What the values of the variable are is something that is stipulated. The stipulation is a description of the propositions that have the variable as their representative. How the description of the terms of the bracketed expression is produced is not essential. We can distinguish three kinds of description: 1.Direct enumeration, in which case we can simply substitute for the variable the constants that are its values; 2. giving a function fx whose values for all values of x are the propositions to be described; 3. giving a formal law that governs the construction of the propositions, in which case the bracketed expression has as its members all the terms of a series of forms.

5.502 So instead of '(———-T) (E, ...)', I write 'N(E)'. N(E) is the negation of all the values of the propositional variable E.

5.503 It is obvious that we can easily express how propositions may be constructed with this operation, and how they may not be constructed with it; so it must be possible to find an exact expression for this.

5.51 If E has only one value, then N(E) = Pp (not p); if it has two values, then N(E) = Pp . Pq. (neither p nor g).

5.511 How can logic—all-embracing logic, which mirrors the world—use such peculiar crotchets and contrivances? Only because they are all connected with one another in an infinitely fine network, the great mirror.

5.512 'Pp' is true if 'p' is false. Therefore, in the proposition 'Pp', when it is true, 'p' is a false proposition. How then can the stroke 'P' make it agree with reality? But in 'Pp' it is not 'P' that negates, it is rather what is common to all the signs of this notation that negate p. That is to say the common rule that governs the construction of 'Pp', 'PPPp', 'Pp C Pp', 'Pp . Pp', etc. etc. (ad inf.). And this common factor mirrors negation.

5.513 We might say that what is common to all symbols that affirm both p and q is the proposition 'p . q'; and that what is common to all symbols that affirm either p or q is the proposition 'p C q'. And similarly we can say that two propositions are opposed to one another if they have nothing in common with one another, and that every proposition has only one negative, since there is only one proposition that lies completely outside it. Thus in Russell's notation too it is manifest that 'q : p C Pp' says the same thing as 'q', that 'p C Pq' says nothing.

5.514 Once a notation has been established, there will be in it a rule governing the construction of all propositions that negate p, a rule governing the construction of all propositions that affirm p, and a rule governing the construction of all propositions that affirm p or q; and so on. These rules are equivalent to the symbols; and in them their sense is mirrored.

5.515 It must be manifest in our symbols that it can only be propositions that are combined with one another by 'C', '.', etc. And this is indeed the case, since the symbol in 'p' and 'q' itself presupposes 'C', 'P', etc. If the sign 'p' in 'p C q' does not stand for a complex sign, then it cannot have sense by itself: but in that case the signs 'p C p', 'p . p', etc., which have the same sense as p, must also lack sense. But if 'p C p' has no sense, then 'p C q' cannot have a sense either.

5.5151 Must the sign of a negative proposition be constructed with that of the positive proposition? Why should it not be possible to express a negative proposition by means of a negative fact? (E.g. suppose that "a' does not stand in a certain relation to 'b'; then this might be used to say that aRb was not the case.) But really even in this case the negative proposition is constructed by an indirect use of the positive. The positive proposition necessarily presupposes the existence of the negative proposition and vice versa.

5.52 If E has as its values all the values of a function fx for all values of x, then N(E) = P(dx) . fx.

5.521 I dissociate the concept all from truth-functions. Frege and Russell introduced generality in association with logical productor logical sum. This made it difficult to understand the propositions '(dx) . fx' and '(x) . fx', in which both ideas are embedded.

5.522 What is peculiar to the generality-sign is first, that it indicates a logical prototype, and secondly, that it gives prominence to constants.

5.523 The generality-sign occurs as an argument.

5.524 If objects are given, then at the same time we are given all objects. If elementary propositions are given, then at the same time all elementary propositions are given.

5.525 It is incorrect to render the proposition '(dx) . fx' in the words, 'fx is possible' as Russell does. The certainty, possibility, or impossibility of a situation is not expressed by a proposition, but by an expression's being a tautology, a proposition with a sense, or a contradiction. The precedent to which we are constantly inclined to appeal must reside in the symbol itself.

5.526 We can describe the world completely by means of fully generalized propositions, i.e. without first correlating any name with a particular object.

5.5261 A fully generalized proposition, like every other proposition, is composite. (This is shown by the fact that in '(dx, O) . Ox' we have to mention 'O' and 's' separately. They both, independently, stand in signifying relations to the world, just as is the case in ungeneralized propositions.) It is a mark of a composite symbol that it has something in common with other symbols.

5.5262 The truth or falsity of every proposition does make some alteration in the general construction of the world. And the range that the totality of elementary propositions leaves open for its construction is exactly the same as that which is delimited by entirely general propositions. (If an elementary proposition is true, that means, at any rate, one more true elementary proposition.)

5.53 Identity of object I express by identity of sign, and not by using a sign for identity. Difference of objects I express by difference of signs.

5.5301 It is self-evident that identity is not a relation between objects. This becomes very clear if one considers, for example, the proposition '(x) : fx . z . x = a'. What this proposition says is simply that only a satisfies the function f, and not that only things that have a certain relation to a satisfy the function, Of course, it might then be said that only a did have this relation to a; but in order to express that, we should need the identity-sign itself.

5.5302 Russell's definition of '=' is inadequate, because according to it we cannot say that two objects have all their properties in common. (Even if this proposition is never correct, it still has sense .)

5.5303 Roughly speaking, to say of two things that they are identical is nonsense, and to say of one thing that it is identical with itself is to say nothing at all.

5.531 Thus I do not write 'f(a, b) . a = b', but 'f(a, a)' (or 'f(b, b)); and not 'f(a,b) . Pa = b', but 'f(a, b)'.

5.532 And analogously I do not write '(dx, y) . f(x, y) . x = y', but '(dx) . f(x, x)'; and not '(dx, y) . f(x, y) . Px = y', but '(dx, y) . f(x, y)'. 5.5321 Thus, for example, instead of '(x) : fx z x = a' we write '(dx) . fx . z : (dx, y) . fx. fy'. And the proposition, 'Only one x satisfies f( )', will read '(dx) . fx : P(dx, y) . fx . fy'.

5.533 The identity-sign, therefore, is not an essential constituent of conceptual notation.

5.534 And now we see that in a correct conceptual notation like 'a = a', 'a = b . b = c . ⊃ a = c', '(x) . x = x', '(∃x) . x = a', etc. cannot even be written down.

5.535 This also disposes of all the problems that were connected with such pseudo-propositions. All the problems that Russell's 'axiom of infinity' brings with it can be solved at this point. What the axiom of infinity is intended to say would express itself in language through the existence of infinitely many names with different meanings.

5.5351 There are certain cases in which one is tempted to use expressions of the form 'a = a' or 'p ⊃ p' and the like. In fact, this happens when one wants to talk about prototypes, e.g. about proposition, thing, etc. Thus in Russell's Principles of Mathematics 'p is a proposition'—which is nonsense--was given the symbolic rendering 'p ⊃ p' and placed as an hypothesis in front of certain propositions in order to exclude from their everything but propositions. (It is nonsense to place the hypothesis 'p ⊃ p' in front of a proposition, in order to ensure that its arguments shall have the right form, if only because with a non-proposition as argument the hypothesis becomes not false but nonsensical, and because arguments of the wrong kind make the proposition itself nonsensical, so that it preserves itself from wrong arguments just as well, or as badly, as the hypothesis without sense that was appended for that purpose.)

5.5352 In the same way people have wanted to express, 'There are no things ', by writing 'P(∃x) . x = x'. But even if this were a proposition, would it not be equally true if in fact 'there were things' but they were not identical with themselves?

5.54 In the general propositional form propositions occur in other propositions only as bases of truth-operations.

5.541 At first sight it looks as if it were also possible for one proposition to occur in another in a different way. Particularly with certain forms of proposition in psychology, such as 'A believes that p is the case' and A has the thought p', etc. For if these are considered superficially, it looks as if the proposition p stood in some kind of relation to an object A. (And in modern theory of knowledge (Russell, Moore, etc.) these propositions have actually been construed in this way.)

5.542 It is clear, however, that 'A believes that p', 'A has the thought p', and 'A says p' are of the form '"p" says p': and this does not involve a correlation of a fact with an object, but rather the correlation of facts by means of the correlation of their objects.

5.5421 This shows too that there is no such thing as the soul—the subject, etc.—as it is conceived in the superficial psychology of the present day. Indeed a composite soul would no longer be a soul.

5.5422 The correct explanation of the form of the proposition, 'A makes the judgement p', must show that it is impossible for a judgement to be a piece of nonsense. (Russell's theory does not satisfy this requirement.)

5.5423 To perceive a complex means to perceive that its constituents are related to one another in such and such a way. This no doubt also explains why there are two possible ways of seeing the figure as a cube; and all similar phenomena. For we really see two different facts. (If I look in the first place at the corners marked a and only glance at the b's, then the a's appear to be in front, and vice versa).

5.55 We now have to answer a priori the question about all the possible forms of elementary propositions. Elementary propositions consist of names. Since, however, we are unable to give the number of names with different meanings, we are also unable to give the composition of elementary propositions.

5.551 Our fundamental principle is that whenever a question can be decided by logic at all it must be possible to decide it without more ado. (And if we get into a position where we have to look at the world for an answer to such a problem, that shows that we are on a completely wrong track.)

5.552 The 'experience' that we need in order to understand logic is not that something or other is the state of things, but that something is : that, however, is not an experience. Logic is prior to every experience—that something is so . It is prior to the question 'How?' not prior to the question 'What?'

5.5521 And if this were not so, how could we apply logic? We might put it in this way: if there would be a logic even if there were no world, how then could there be a logic given that there is a world?

5.553 Russell said that there were simple relations between different numbers of things (individuals). But between what numbers? And how is this supposed to be decided?—By experience? (There is no pre-eminent number.)

5.554 It would be completely arbitrary to give any specific form.

5.5541 It is supposed to be possible to answer a priori the question whether I can get into a position in which I need the sign for a 27-termed relation in order to signify something.

5.5542 But is it really legitimate even to ask such a question? Can we set up a form of sign without knowing whether anything can correspond to it? Does it make sense to ask what there must be in order that something can be the case?

5.555 Clearly we have some concept of elementary propositions quite apart from their particular logical forms. But when there is a system by which we can create symbols, the system is what is important for logic and not the individual symbols. And anyway, is it really possible that in logic I should have to deal with forms that I can invent? What I have to deal with must be that which makes it possible for me to invent them.

5.556 There cannot be a hierarchy of the forms of elementary propositions. We can foresee only what we ourselves construct.

5.5561 Empirical reality is limited by the totality of objects. The limit also makes itself manifest in the totality of elementary propositions. Hierarchies are and must be independent of reality.

5.5562 If we know on purely logical grounds that there must be elementary propositions, then everyone who understands propositions in their C form must know It.

5.5563 In fact, all the propositions of our everyday language, just as they stand, are in perfect logical order.—That utterly simple thing, which we have to formulate here, is not a likeness of the truth, but the truth itself in its entirety. (Our problems are not abstract, but perhaps the most concrete that there are.)

5.557 The application of logic decides what elementary propositions there are. What belongs to its application, logic cannot anticipate. It is clear that logic must not clash with its application. But logic has to be in contact with its application. Therefore logic and its application must not overlap.

5.5571 If I cannot say a priori what elementary propositions there are, then the attempt to do so must lead to obvious nonsense. 5.6 The limits of my language mean the limits of my world.

5.61 Logic pervades the world: the limits of the world are also its limits. So we cannot say in logic, 'The world has this in it, and this, but not that.' For that would appear to presuppose that we were excluding certain possibilities, and this cannot be the case, since it would require that logic should go beyond the limits of the world; for only in that way could it view those limits from the other side as well. We cannot think what we cannot think; so what we cannot think we cannot say either.

5.62 This remark provides the key to the problem, how much truth there is in solipsism. For what the solipsist means is quite correct; only it cannot be said , but makes itself manifest. The world is my world: this is manifest in the fact that the limits of language (of that language which alone I understand) mean the limits of my world.

5.621 The world and life are one.

5.63 I am my world. (The microcosm.)

5.631 There is no such thing as the subject that thinks or entertains ideas. If I wrote a book called The World as I found it , I should have to include a report on my body, and should have to say which parts were subordinate to my will, and which were not, etc., this being a method of isolating the subject, or rather of showing that in an important sense there is no subject; for it alone could not be mentioned in that book.—

5.632 The subject does not belong to the world: rather, it is a limit of the world.

5.633 Where in the world is a metaphysical subject to be found? You will say that this is exactly like the case of the eye and the visual field. But really you do not see the eye. And nothing in the visual field allows you to infer that it is seen by an eye.

5.6331 For the form of the visual field is surely not like this

5.634 This is connected with the fact that no part of our experience is at the same time a priori. Whatever we see could be other than it is. Whatever we can describe at all could be other than it is. There is no a priori order of things.

5.64 Here it can be seen that solipsism, when its implications are followed out strictly, coincides with pure realism. The self of solipsism shrinks to a point without extension, and there remains the reality co-ordinated with it.

5.641 Thus there really is a sense in which philosophy can talk about the self in a non-psychological way. What brings the self into philosophy is the fact that 'the world is my world'. The philosophical self is not the human being, not the human body, or the human soul, with which psychology deals, but rather the metaphysical subject, the limit of the world—not a part of it.

6 The general form of a truth-function is [p, E, N(E)]. This is the general form of a proposition.

6.001 What this says is just that every proposition is a result of successive applications to elementary propositions of the operation N(E)

6.002 If we are given the general form according to which propositions are constructed, then with it we are also given the general form according to which one proposition can be generated out of another by means of an operation.

6.01 Therefore the general form of an operation /'(n) is [E, N(E)] ' (n) (= [n, E, N(E)]). This is the most general form of transition from one proposition to another.

6.02 And this is how we arrive at numbers. I give the following definitions x = /0x Def., /'/v'x = /v+1'x Def. So, in accordance with these rules, which deal with signs, we write the series x, /'x, /'/'x, /'/'/'x, ... , in the following way /0'x, /0+1'x, /0+1+1'x, /0+1+1+1'x, ... Therefore, instead of '[x, E, /'E]', I write '[/0'x, /v'x, /v+1'x]'. And I give the following definitions 0 + 1 = 1 Def., 0 + 1 + 1 = 2 Def., 0 + 1 + 1 +1 = 3 Def., (and so on).

6.021 A number is the exponent of an operation.

6.022 The concept of number is simply what is common to all numbers, the general form of a number. The concept of number is the variable number. And the concept of numerical equality is the general form of all particular cases of numerical equality.

6.03 The general form of an integer is [0, E, E +1].

6.031 The theory of classes is completely superfluous in mathematics. This is connected with the fact that the generality required in mathematics is not accidental generality.

6.1 The propositions of logic are tautologies.

6.11 Therefore the propositions of logic say nothing. (They are the analytic propositions.)

6.111 All theories that make a proposition of logic appear to have content are false. One might think, for example, that the words 'true' and 'false' signified two properties among other properties, and then it would seem to be a remarkable fact that every proposition possessed one of these properties. On this theory it seems to be anything but obvious, just as, for instance, the proposition, 'All roses are either yellow or red', would not sound obvious even if it were true. Indeed, the logical proposition acquires all the characteristics of a proposition of natural science and this is the sure sign that it has been construed wrongly.

6.112 The correct explanation of the propositions of logic must assign to them a unique status among all propositions.

6.113 It is the peculiar mark of logical propositions that one can recognize that they are true from the symbol alone, and this fact contains in itself the whole philosophy of logic. And so too it is a very important fact that the truth or falsity of non-logical propositions cannot be recognized from the propositions alone.

6.12 The fact that the propositions of logic are tautologies shows the formal—logical—properties of language and the world. The fact that a tautology is yielded by this particular way of connecting its constituents characterizes the logic of its constituents. If propositions are to yield a tautology when they are connected in a certain way, they must have certain structural properties. So their yielding a tautology when combined in this shows that they possess these structural properties.

6.1201 For example, the fact that the propositions 'p' and 'Pp' in the combination '(p . Pp)' yield a tautology shows that they contradict one another. The fact that the propositions 'p z q', 'p', and 'q', combined with one another in the form '(p z q) . (p) :z: (q)', yield a tautology shows that q follows from p and p z q. The fact that '(x) . fxx :z: fa' is a tautology shows that fa follows from (x) . fx. Etc. etc.

6.1202 It is clear that one could achieve the same purpose by using contradictions instead of tautologies.

6.1203 In order to recognize an expression as a tautology, in cases where no generality-sign occurs in it, one can employ the following intuitive method: instead of 'p', 'q', 'r', etc. I write 'TpF', 'TqF', 'TrF', etc. Truth-combinations I express by means of brackets, e.g. and I use lines to express the correlation of the truth or falsity of the whole proposition with the truth-combinations of its truth-arguments, in the following way So this sign, for instance, would represent the proposition p z q. Now, by way of example, I wish to examine the proposition P(p .Pp) (the law of contradiction) in order to determine whether it is a tautology. In our notation the form 'PE' is written as and the form 'E . n' as Hence the proposition P(p . Pp). reads as follows If we here substitute 'p' for 'q' and examine how the outermost T and F are connected with the innermost ones, the result will be that the truth of the whole proposition is correlated with all the truth-combinations of its argument, and its falsity with none of the truth-combinations.

6.121 The propositions of logic demonstrate the logical properties of propositions by combining them so as to form propositions that say nothing. This method could also be called a zero-method. In a logical proposition, propositions are brought into equilibrium with one another, and the state of equilibrium then indicates what the logical constitution of these propositions must be.

6.122 It follows from this that we can actually do without logical propositions; for in a suitable notation we can in fact recognize the formal properties of propositions by mere inspection of the propositions themselves.

6.1221 If, for example, two propositions 'p' and 'q' in the combination 'p z q' yield a tautology, then it is clear that q follows from p. For example, we see from the two propositions themselves that 'q' follows from 'p z q . p', but it is also possible to show it in this way: we combine them to form 'p z q . p :z: q', and then show that this is a tautology.

6.1222 This throws some light on the question why logical propositions cannot be confirmed by experience any more than they can be refuted by it. Not only must a proposition of logic be irrefutable by any possible experience, but it must also be unconfirmable by any possible experience.

6.1223 Now it becomes clear why people have often felt as if it were for us to 'postulate' the 'truths of logic'. The reason is that we can postulate them in so far as we can postulate an adequate notation.

6.1224 It also becomes clear now why logic was called the theory of forms and of inference.

6.123 Clearly the laws of logic cannot in their turn be subject to laws of logic. (There is not, as Russell thought, a special law of contradiction for each 'type'; one law is enough, since it is not applied to itself.)

6.1231 The mark of a logical proposition is not general validity. To be general means no more than to be accidentally valid for all things. An ungeneralized proposition can be tautological just as well as a generalized one.

6.1232 The general validity of logic might be called essential, in contrast with the accidental general validity of such propositions as 'All men are mortal'. Propositions like Russell's 'axiom of reducibility' are not logical propositions, and this explains our feeling that, even if they were true, their truth could only be the result of a fortunate accident.

6.1233 It is possible to imagine a world in which the axiom of reducibility is not valid. It is clear, however, that logic has nothing to do with the question whether our world really is like that or not.

6.124 The propositions of logic describe the scaffolding of the world, or rather they represent it. They have no 'subject-matter'.

They presuppose that names have meaning and elementary propositions sense; and that is their connexion with the world. It is clear that something about the world must be indicated by the fact that certain combinations of symbols—whose essence involves the possession of a determinate character—are tautologies. This contains the decisive point. We have said that some things are arbitrary in the symbols that we use and that some things are not. In logic it is only the latter that express: but that means that logic is not a field in which we express what we wish with the help of signs, but rather one in which the nature of the absolutely necessary signs speaks for itself. If we know the logical syntax of any sign-language, then we have already been given all the propositions of logic.

6.125 It is possible—indeed possible even according to the old conception of logic—to give in advance a description of all 'true' logical propositions.

6.1251 Hence there can never be surprises in logic.

6.126 One can calculate whether a proposition belongs to logic, by calculating the logical properties of the symbol. And this is what we do when we 'prove' a logical proposition. For, without bothering about sense or meaning, we construct the logical proposition out of others using only rules that deal with signs . The proof of logical propositions consists in the following process: we produce them out of other logical propositions by successively applying certain operations that always generate further tautologies out of the initial ones. (And in fact only tautologies follow from a tautology.) Of course this way of showing that the propositions of logic are tautologies is not at all essential to logic, if only because the propositions from which the proof starts must show without any proof that they are tautologies.

6.1261 In logic process and result are equivalent. (Hence the absence of surprise.)

6.1262 Proof in logic is merely a mechanical expedient to facilitate the recognition of tautologies in complicated cases.

6.1263 Indeed, it would be altogether too remarkable if a proposition that had sense could be proved logically from others, and so too could a logical proposition. It is clear from the start that a logical proof of a proposition that has sense and a proof in logic must be two entirely different things.

6.1264 A proposition that has sense states something, which is shown by its proof to be so. In logic every proposition is the form of a proof. Every proposition of logic is a modus ponens represented in signs. (And one cannot express the modus ponens by means of a proposition.)

6.1265 It is always possible to construe logic in such a way that every proposition is its own proof.

6.127 All the propositions of logic are of equal status: it is not the case that some of them are essentially derived propositions. Every tautology itself shows that it is a tautology.

6.1271 It is clear that the number of the 'primitive propositions of logic' is arbitrary, since one could derive logic from a single primitive proposition, e.g. by simply constructing the logical product of Frege's primitive propositions. (Frege would perhaps say that we should then no longer have an immediately self-evident primitive proposition. But it is remarkable that a thinker as rigorous as Frege appealed to the degree of self-evidence as the criterion of a logical proposition.)

6.13 Logic is not a body of doctrine, but a mirror-image of the world. Logic is transcendental.

6.2 Mathematics is a logical method. The propositions of mathematics are equations, and therefore pseudo-propositions.

6.21 A proposition of mathematics does not express a thought.

6.211 Indeed in real life a mathematical proposition is never what we want. Rather, we make use of mathematical propositions only in inferences from propositions that do not belong to mathematics to others that likewise do not belong to mathematics. (In philosophy the question, 'What do we actually use this word or this proposition for?' repeatedly leads to valuable insights.)

6.22 The logic of the world, which is shown in tautologies by the propositions of logic, is shown in equations by mathematics.

6.23 If two expressions are combined by means of the sign of equality, that means that they can be substituted for one another. But it must be manifest in the two expressions themselves whether this is the case or not. When two expressions can be substituted for one another, that characterizes their logical form.

6.231 It is a property of affirmation that it can be construed as double negation. It is a property of '1 + 1 + 1 + 1' that it can be construed as '(1 + 1) + (1 + 1)'.

6.232 Frege says that the two expressions have the same meaning but different senses. But the essential point about an equation is that it is not necessary in order to show that the two expressions connected by the sign of equality have the same meaning, since this can be seen from the two expressions themselves.

6.2321 And the possibility of proving the propositions of mathematics means simply that their correctness can be perceived without its being necessary that what they express should itself be compared with the facts in order to determine its correctness.

6.2322 It is impossible to assert the identity of meaning of two expressions. For in order to be able to assert anything about their meaning, I must know their meaning, and I cannot know their meaning without knowing whether what they mean is the same or different.

6.2323 An equation merely marks the point of view from which I consider the two expressions: it marks their equivalence in meaning.

6.233 The question whether intuition is needed for the solution of mathematical problems must be given the answer that in this case language itself provides the necessary intuition.

6.2331 The process of calculating serves to bring about that intuition. Calculation is not an experiment.

6.234 Mathematics is a method of logic.

6.2341 It is the essential characteristic of mathematical method that it employs equations. For it is because of this method that every proposition of mathematics must go without saying.

6.24 The method by which mathematics arrives at its equations is the method of substitution. For equations express the substitutability of two expressions and, starting from a number of equations, we advance to new equations by substituting different expressions in accordance with the equations.

6.241 Thus the proof of the proposition 2 t 2 = 4 runs as follows: (/v)n'x = /v x u'x Def., /2 x 2'x = (/2)2'x = (/2)1 + 1'x = /2' /2'x = /1 + 1'/1 + 1'x = (/'/)'(/'/)'x = /'/'/'/'x = /1 + 1 + 1 + 1'x = /4'x.

6.3 The
exploration of logic means the exploration of everything that is subject to
law . And outside logic everything is accidental.

6.31 The so-called law of induction cannot possibly be a law of logic, since it is obviously a proposition with sense.—-Nor, therefore, can it be an a priori law.

6.32 The law of causality is not a law but the form of a law.

6.321 'Law of causality'—that is a general name. And just as in mechanics, for example, there are 'minimum-principles', such as the law of least action, so too in physics there are causal laws, laws of the causal form.

6.3211 Indeed people even surmised that there must be a 'law of least action' before they knew exactly how it went. (Here, as always, what is certain a priori proves to be something purely logical.)

6.33 We do not have an a priori belief in a law of conservation, but rather a priori knowledge of the possibility of a logical form.

6.34 All such propositions, including the principle of sufficient reason, tile laws of continuity in nature and of least effort in nature, etc. etc.—all these are a priori insights about the forms in which the propositions of science can be cast.

6.341 Newtonian mechanics, for example, imposes a unified form on the description of the world. Let us imagine a white surface with irregular black spots on it. We then say that whatever kind of picture these make, I can always approximate as closely as I wish to the description of it by covering the surface with a sufficiently fine square mesh, and then saying of every square whether it is black or white. In this way I shall have imposed a unified form on the description of the surface. The form is optional, since I could have achieved the same result by using a net with a triangular or hexagonal mesh. Possibly the use of a triangular mesh would have made the description simpler: that is to say, it might be that we could describe the surface more accurately with a coarse triangular mesh than with a fine square mesh (or conversely), and so on. The different nets correspond to different systems for describing the world. Mechanics determines one form of description of the world by saying that all propositions used in the description of the world must be obtained in a given way from a given set of propositions—the axioms of mechanics. It thus supplies the bricks for building the edifice of science, and it says, 'Any building that you want to erect, whatever it may be, must somehow be constructed with these bricks, and with these alone.' (Just as with the number-system we must be able to write down any number we wish, so with the system of mechanics we must be able to write down any proposition of physics that we wish.)

6.342 And now we can see the relative position of logic and mechanics. (The net might also consist of more than one kind of mesh: e.g. we could use both triangles and hexagons.) The possibility of describing a picture like the one mentioned above with a net of a given form tells us nothing about the picture. (For that is true of all such pictures.) But what does characterize the picture is that it can be described completely by a particular net with a particular size of mesh. Similarly the possibility of describing the world by means of Newtonian mechanics tells us nothing about the world: but what does tell us something about it is the precise way in which it is possible to describe it by these means. We are also told something about the world by the fact that it can be described more simply with one system of mechanics than with another.

6.343 Mechanics is an attempt to construct according to a single plan all the true propositions that we need for the description of the world.

6.3431 The laws of physics, with all their logical apparatus, still speak, however indirectly, about the objects of the world.

6.3432 We ought not to forget that any description of the world by means of mechanics will be of the completely general kind. For example, it will never mention particular point-masses: it will only talk about any whatsoever.

6.35 Although the spots in our picture are geometrical figures, nevertheless geometry can obviously say nothing at all about their actual form and position. The network, however, is purely geometrical; all its properties can be given a priori. Laws like the principle of sufficient reason, etc. are about the net and not about what the net describes.

6.36 If there were a law of causality, it might be put in the following way: There are laws of nature. But of course that cannot be said: it makes itself manifest.

6.361 One might say, using Hertt:'s terminology, that only connexions that are subject to law are thinkable.

6.3611 We cannot compare a process with 'the passage of time'—there is no such thing—but only with another process (such as the working of a chronometer). Hence we can describe the lapse of time only by relying on some other process. Something exactly analogous applies to space: e.g. when people say that neither of two events (which exclude one another) can occur, because there is nothing to cause the one to occur rather than the other, it is really a matter of our being unable to describe one of the two events unless there is some sort of asymmetry to be found. And if such an asymmetry is to be found, we can regard it as the cause of the occurrence of the one and the non-occurrence of the other.

6.36111 Kant's problem about the right hand and the left hand, which cannot be made to coincide, exists even in two dimensions. Indeed, it exists in one-dimensional space in which the two congruent figures, a and b, cannot be made to coincide unless they are moved out of this space. The right hand and the left hand are in fact completely congruent. It is quite irrelevant that they cannot be made to coincide. A right-hand glove could be put on the left hand, if it could be turned round in four-dimensional space.

6.362 What can be described can happen too: and what the law of causality is meant to exclude cannot even be described.

6.363 The procedure of induction consists in accepting as true the simplest law that can be reconciled with our experiences.

6.3631 This procedure, however, has no logical justification but only a psychological one. It is clear that there are no grounds for believing that the simplest eventuality will in fact be realized.

6.36311 It is an hypothesis that the sun will rise tomorrow: and this means that we do not know whether it will rise.

6.37 There is no compulsion making one thing happen because another has happened. The only necessity that exists is logical necessity.

6.371 The whole modern conception of the world is founded on the illusion that the so-called laws of nature are the explanations of natural phenomena.

6.372 Thus people today stop at the laws of nature, treating them as something inviolable, just as God and Fate were treated in past ages. And in fact both are right and both wrong: though the view of the ancients is clearer in so far as they have a clear and acknowledged terminus, while the modern system tries to make it look as if everything were explained.

6.373 The world is independent of my will.

6.374 Even if all that we wish for were to happen, still this would only be a favour granted by fate, so to speak: for there is no logical connexion between the will and the world, which would guarantee it, and the supposed physical connexion itself is surely not something that we could will.

6.375 Just as the only necessity that exists is logical necessity, so too the only impossibility that exists is logical impossibility.

6.3751 For example, the simultaneous presence of two colours at the same place in the visual field is impossible, in fact logically impossible, since it is ruled out by the logical structure of colour. Let us think how this contradiction appears in physics: more or less as follows—a particle cannot have two velocities at the same time; that is to say, it cannot be in two places at the same time; that is to say, particles that are in different places at the same time cannot be identical. (It is clear that the logical product of two elementary propositions can neither be a tautology nor a contradiction. The statement that a point in the visual field has two different colours at the same time is a contradiction.)

6.4 All propositions are of equal value.

6.41 The sense of the world must lie outside the world. In the world everything is as it is, and everything happens as it does happen: in it no value exists—and if it did exist, it would have no value. If there is any value that does have value, it must lie outside the whole sphere of what happens and is the case. For all that happens and is the case is accidental. What makes it non-accidental cannot lie within the world, since if it did it would itself be accidental. It must lie outside the world.

6.42 So too it is impossible for there to be propositions of ethics. Propositions can express nothing that is higher.

6.421 It is clear that ethics cannot be put into words. Ethics is transcendental. (Ethics and aesthetics are one and the same.)

6.422 When an ethical law of the form, 'Thou shalt ...' is laid down, one's first thought is, 'And what if I do, not do it?' It is clear, however, that ethics has nothing to do with punishment and reward in the usual sense of the terms. So our question about the consequences of an action must be unimportant.—At least those consequences should not be events. For there must be something right about the question we posed. There must indeed be some kind of ethical reward and ethical punishment, but they must reside in the action itself. (And it is also clear that the reward must be something pleasant and the punishment something unpleasant.)

6.423 It is impossible to speak about the will in so far as it is the subject of ethical attributes. And the will as a phenomenon is of interest only to psychology.

6.43 If the good or bad exercise of the will does alter the world, it can alter only the limits of the world, not the facts—not what can be expressed by means of language. In short the effect must be that it becomes an altogether different world. It must, so to speak, wax and wane as a whole. The world of the happy man is a different one from that of the unhappy man.

6.431 So too at death the world does not alter, but comes to an end.

6.4311 Death is not an event in life: we do not live to experience death. If we take eternity to mean not infinite temporal duration but timelessness, then eternal life belongs to those who live in the present. Our life has no end in just the way in which our visual field has no limits.

6.4312 Not only is there no guarantee of the temporal immortality of the human soul, that is to say of its eternal survival after death; but, in any case, this assumption completely fails to accomplish the purpose for which it has always been intended. Or is some riddle solved by my surviving for ever? Is not this eternal life itself as much of a riddle as our present life? The solution of the riddle of life in space and time lies outside space and time. (It is certainly not the solution of any problems of natural science that is required.)

6.432 How things are in the world is a matter of complete indifference for what is higher. God does not reveal himself in the world.

6.4321 The facts all contribute only to setting the problem, not to its solution.

6.44 It is not how things are in the world that is mystical, but that it exists.

6.45 To view the world sub specie aeterni is to view it as a whole—a limited whole. Feeling the world as a limited whole—it is this that is mystical.

6.5 When the answer cannot be put into words, neither can the question be put into words. The riddle does not exist. If a question can be framed at all, it is also possible to answer it.

6.51 Scepticism is not irrefutable, but obviously nonsensical, when it tries to raise doubts where no questions can be asked. For doubt can exist only where a question exists, a question only where an answer exists, and an answer only where something can be said.

6.52 We feel that even when all possible scientific questions have been answered, the problems of life remain completely untouched. Of course there are then no questions left, and this itself is the answer.

6.521 The solution of the problem of life is seen in the vanishing of the problem. (Is not this the reason why those who have found after a long period of doubt that the sense of life became clear to them have then been unable to say what constituted that sense?)

6.522 There are, indeed, things that cannot be put into words. They make themselves manifest. They are what is mystical.

6.53 The correct method in philosophy would really be the following: to say nothing except what can be said, i.e. propositions of natural science—i.e. something that has nothing to do with philosophy—and then, whenever someone else wanted to say something metaphysical, to demonstrate to him that he had failed to give a meaning to certain signs in his propositions. Although it would not be satisfying to the other person—he would not have the feeling that we were teaching him philosophy—this method would be the only strictly correct one.

6.54 My propositions are elucidatory in this way: he who understands me finally recognizes them as senseless, when he has climbed out through them, on them, over them. (He must so to speak throw away the ladder, after he has climbed up on it.)

7 What we cannot speak about we must pass over in silence.7 What we cannot speak about we must pass over in silence.

www.ingramcontent.com/pod-product-compliance
Lightning Source LLC
Chambersburg PA
CBHW020359170426
43200CB00005B/228